CORPORATE
ENVIRONMENTAL
MANAGEMENT

CORPORATE ENVIRONMENTAL MANAGEMENT

JOHN DARABARIS, P.E., C.P.A.

CRC Press
Taylor & Francis Group
Boca Raton   London   New York

CRC Press is an imprint of the
Taylor & Francis Group, an **informa** business

CRC Press
Taylor & Francis Group
6000 Broken Sound Parkway NW, Suite 300
Boca Raton, FL 33487-2742

© 2008 by Taylor & Francis Group, LLC
CRC Press is an imprint of Taylor & Francis Group, an Informa business

International Standard Book Number-13: 978-1-4200-5546-7 (Hardcover)

Library of Congress Cataloging-in-Publication Data

Darabaris, John.
    Corporate environmental management / John Darabaris.
        p. cm.
    Includes bibliographical references and index.
    ISBN 978-1-4200-5546-7 (alk. paper)
        1. Environmental management--Industrial applications. 2. Environmental risk assessment--Industrial applications. 3. Social responsibility of business--Environmental aspects. 4. Industrial management--Environmental aspects. I. Title.

GE300.D37 2008
658.4'083--dc22   2007016722

Visit the Taylor & Francis Web site at
http://www.taylorandfrancis.com

and the CRC Press Web site at
http://www.crcpress.com

Table of Contents

Preface ix
About the Author xi
List of Illustrations xiii

1   Introduction **1**
     Reference 7

2   Environmental Management Assessment **9**

3   Lines of Inquiry **13**

4   Assessment Model and Analytical Framework **17**
     Assessment Model 17
     Analytical Framework 19
     Program vs. Project Management 20
     Environmental Risk Management 22

5   Internal Survey **27**
     Corporate Headquarters 27
     Operations 29

6   Corporate Commitment **31**
     Senior Management Commitment 31
     Corporate Environmental Policy 33
          Responsible Care 34
          The CERES Principles 35
          GEMI and the ICC Charter 35
          United Nations Environment Programmes' Financial Institutions
               Initiative on the Environment 36
          Environmental Banking Association 36
          World Business Council for Sustainable Development 36
          ISO 14000 37
     Performance Track Corporate Leaders 40
     Strategic Environmental Planning 40
     References 42

7   Communication **43**
     External Communication/Public Relations 43

Internal Communication 46
Issue Management..48

8   Functional Implementation...51
Organization and Staffing..51
Internal Integration...52
Operational Systems...53
     Pollution Prevention Opportunities 55
Environmental Risk Management...60
Waste Minimization Programs...61
The Emission Reduction Program...63

9   Measurement Systems..65
Audit Program 65
     Site Tours 68
     Root Cause Analysis...69
Environmental Management Information System (MIS).............70
Environmental Cost Management 72
Environmental Asset Management..75
Environmental Financial Management.....................................77
     Conventional Commercial Loan 79
     Supplier Financing...79
     Commercial Paper 80
     Bond Financing..80
     Private Placement Debt...81
     Environmental Capital Equipment Leasing....................81
     Environmental Related Preferred Stock 82
     Master Limited Partnerships..82
     Research and Development...82
References..86

10   Benchmark Survey 87
Environmental Management-Based Benchmarking Survey Approach........89
Technical-Based Benchmarking Survey Approach......................91
References..96

11   External Survey 99
External Scan..99
Assessing Global Impacts—Sustainable Development................99
Project Life Cycle Analysis...103
Activist Group Alliances...105
Independent Technology Scan...107

12   Natural Resource Damage Assessment—Proactive Strategies....109
Early Recognition of Contamination Stage...............................113
Agreement or Settlement Stage 115
Cleanup Stage...115
Pursuit of a PRP Claim Stage..115
References...115

13   Environmental Risk Assessment Issues 117
General Discussion 117
  Health Assessments 117
  Ecological Risk Assessment 120
Population Risk Analysis 122
References 123

14   Emergency Response Analysis 125
Contingency Plans 126
Up-the-Stack Emergencies 127
General Emergency Management Concepts 130
References 134

15   Corporate Health and Safety System....................135
Establishing Hazard and Safety Control Measures 136
  Inorganic Chemicals 137
  Organic Compounds 138
  Operational Chemicals/Hazard Communication Program 138
  Personal Protective Equipment 139
    Respiratory Protection 139
    Levels of Protection..140
    Monitoring and Medical Surveillance 143
  Site Control of Work Zones 144
    Exclusion Zone (EZ) 144
References 144

16   Environmental Risk Management at Banking Institutions 147
Practices for the Commercial Banking Community 149
  EPA Lender Liability Rule 151
  Post-Commitment Practices for Debt Transactions 152
Practices for the Equity Banking Community 154
Integrating Environmental and Financial Performance..........157
References 164

17   Global Warming..167
Existing Market and Potential Revenue 168
Brokers...169
Global Climate Profile 169
Global Climate Summary 172
References 174

18   Assessment of International Trends 175
OECD/EIRIS Study Results 175
Survey on the State of Global Environmental and Social Reporting 186
Emerging China and India Environmental Issues.................187
Kyoto Protocol Debates 188
State of the U.S. Environmental Policies 189

      "Greenwash" versus "Green Machine" Debate 189
           ExxonMobil 190
           Wal-Mart 190
      References 190

19   Summary 193
      References 196

Index 197

Preface

The goal of corporate environmental management is threefold:

- To provide a basis for the independent assessment of environmental management that marries the various standardized approaches for measuring components (e.g., environmental audits for facilities, ISO 14000 compliance) with the larger and more sophisticated goals of overall corporate management objectives;
- To provide summary inputs regarding various global environmental management initiatives and developments that may be of interest to the target audience (with full recognition that this is a moving target); and
- To provide technical and management insights and suggestions to aid environmental management professionals and their corporate management structure in their development and implementation of initiatives, as well as providing interested investment and stakeholder communities a basis for independent evaluation.

The target audience is also threefold:

- To provide senior management and boards of directors a concise, independent approach to assessing their respective environmental management programs from a senior executive perspective;
- To provide the investment community with an independent perspective for evaluating corporate environmental management performance of their portfolio (and prospective portfolio) as well as updates on the emerging environmental stance within the investment community and its regulatory bodies (e.g., SEC); and finally,
- To provide the environmental management community itself with suggestions and implementation strategies for basic issues such as health and safety, clean air, clean water, CERCLA, and NRDA, as well as emerging issues such as risk management, conservation science, and sustainable development reporting.

Last, corporate environmental management has a fourth goal and target audience—to provide a sound crossover primer for the academic community, providing a science and regulatory perspective to the MBA community, and a management perspective to the environmental science/engineering graduate community.

About the Author

John Darabaris, currently in a management capacity, was formerly a division vice president with Kearney/Centaur, where he conducted numerous corporate environmental management assessments and benchmarking studies as well as "best practice" studies. He is an experienced environmental management professional knowledgeable in both environmental management and regulatory strategy as related to complex, sophisticated, industry environmental activities. Possessing both a professional engineer (PE) license and a non-practicing certified public accountant (CPA) certificate, he marries both engineering and management perspectives to the impacts of corporate environmental management and related regulatory strategy development.

With a background that combines graduate degrees in geologic engineering (MS, University of Missouri at Rolla) and finance (MBA, Columbia University, New York), Darabaris provides unique insights into the breadth of technical, regulatory, and management issues that corporate environmental management teams face in today's complex environmental corrective-action management world.

In recognition of his achievements, Darabaris was awarded an honorary professional development degree from the University of Missouri at Rolla and a commendation from the U.S. Army Corps of Engineers' Omaha office.

This is Darabaris' second book. His first, *Macroengineering: An Environmental Management Restoration Management Process* (CRC Press, 2006), is currently on the market and has sold worldwide in over 20 countries and over a dozen universities.

List of Illustrations

Exhibit 1. New environmental problems and the environmental competition of companies.
Exhibit 2. Evolution and assessment of environmental management.
Exhibit 3. Business consequences of the "efficient compliance" approach.
Exhibit 4. Few environmental restoration chains are effectively integrated.
Exhibit 5. Integrated environmental management programs.
Exhibit 6. Components of an SC&A environmental management assessment.
Exhibit 7. Elements of an external survey.
Exhibit 8. Environmental management assessments: Stages of Excellence.
Exhibit 9. Principal analytical criteria.
Exhibit 10. Program versus project management analysis.
Exhibit 11. Interwoven risk perspectives.
Exhibit 12. Perceived risk reflects public attitude.
Exhibit 13. Environmental management risk review process.
Exhibit 14. Environmental management regulatory model.
Exhibit 15. Positioning the environmental management role.
Exhibit 16. Internal survey, examples of breadth versus depth.
Exhibit 17. Risk management "audit" versus "risk" management cultures.
Exhibit 18. Strategic environmental planning model.
Exhibit 19. External communication links.
Exhibit 20. Straw model communications plan matrix.
Exhibit 21. Internal communication links.
Exhibit 22. Instituting ongoing feedback and learning.
Exhibit 23. Examples of functional implementation with other corporate groups.
Exhibit 24. Disruptive environmental technologies: A need for leadership.
Exhibit 25. Examples of process or equipment modifications enhancements.
Exhibit 26. Examples of potential maintenance enhancements.
Exhibit 27. Examples of potential waste segregation and separation enhancements.
Exhibit 28. Examples of recycling and potential material substitution enhancements.
Exhibit 29. Pathways to environmental risk management.
Exhibit 30. Functional implementation criteria model.
Exhibit 31. Metrics Development Wheel.
Exhibit 32. Examples of potential environmental management information system (MIS) elements.
Exhibit 33. The EHS cost pyramid.
Exhibit 34. The Ricoh approach to environmental accounting.
Exhibit 35. Strategic management of EHS investments.
Exhibit 36. Example of an environmental asset value chain.
Exhibit 37. A capital value approach to establishing value of performance.

Exhibit 38. A risk-based decision approach to valuing performance assets.
Exhibit 39. Using stakeholder satisfaction to value excitement assets.
Exhibit 40. Establishing the competitive position of assets.
Exhibit 41. A portfolio approach to valuing EHS assets.
Exhibit 42. EcoVALUE'21™ variables.
Exhibit 43. Benchmarking environmental practice and performance.
Exhibit 44. Example of process for benchmarking environmental management structure and effectiveness.
Exhibit 45. Difference between internal and external views gained from benchmarking.
Exhibit 46. Criteria for environmental management success.
Exhibit 47. Stages of the environmental management success model.
Exhibit 48. Performance parameter examples.
Exhibit 49. Sample format of a benchmarking questionnaire layout.
Exhibit 50. Sample of a display plot format.
Exhibit 51. List of the EPA's Office of Compliance Industry Sector Notebooks.
Exhibit 52. 1993 Pollutant Releases (short tons/year), from the Petroleum Refining Industry Sector Notebook, September 1995.
Exhibit 53. Summary of 1993 Toxics Release Inventory (TRI) data: Releases and transfers by industry, from the Petroleum Refinery Industry Sector Notebooks, September 1995.
Exhibit 54. 1993 Toxics Release Inventory (TRI) data for selected industries (Source: Petroleum Refinery Industry Sector Notebooks, September 1995).
Exhibit 55. Potential external scan participants.
Exhibit 56. Assessing global impacts.
Exhibit 57. Forces driving sustainable development.
Exhibit 58. Scenario for building the business environment.
Exhibit 59. Environmental management audit technology success criteria.
Exhibit 60. Framework for natural resource damage (NRD) claims.
Exhibit 61. Natural resource damage (NRD) assessment process (as established by 43 CFR Part II).
Exhibit 62. Prototypical examples of natural resource damage (NRD) claim expansion impact.
Exhibit 63. Types of environmental restoration sites.
Exhibit 64. Ten technical defense tips: Areas of potentially "unrealistic expectations" on which to focus.
Exhibit 65. Dose response curve (dose, arbitrary units, logarithmic scale). Routes of entry: inhalation, ingestion, absorption, injection.
Exhibit 66. UEL/LEL example for gasoline.
Exhibit 67. Levels of protection: Level A, Level B, Level C, and Level D.
Exhibit 68. Environmental strategies: A corporate view.
Exhibit 69. Global climate change lexicon.
Exhibit 70. Six Kyoto greenhouse gases (GHG).
Exhibit 71. Share of enterprises that publish environmental policy statements.
Exhibit 72. Companies in FTSE All-World Developed Index, by nationality and sector.
Exhibit 73. Contents of environmental policy statements, all sectors.
Exhibit 74. Signatories to Voluntary Initiatives.
Exhibit 75. Share of enterprises that have implemented environmental policy statements.
Exhibit 76. Share of enterprises that undertake environmental performance reports.
Exhibit 77. Nature of companies' environmental performance reports (percentage share of companies that issue EPR/EPS).

Exhibit 78. **Evidence of the presence of an occupational health and safety system (percentage share, by country or region).**
Exhibit 79. **Different perspectives of environmental management performance.**
Exhibit 80. **Sustainable development.**
Exhibit 81. **Understanding sustainable development.**
Exhibit 82. **Recognize the emotional cycle for environmental management change.**

Chapter 1
Introduction

There is a new era of global environmental factors that needs to be addressed via sound Corporate Environmental Management. Environmental problems are a key area of concern for the global community in the new 21st century. A sustainable global community is emerging that is geared toward preserving limited resources and the natural ecology, but to do so requires resource conservation and the reduction of environmental pollution loads across all human activities.

This text focuses on the emerging relationship between corporate management and the environment as a new era arrives on the scene, where environmental factors increasingly play a key role in corporate competition and generate a need for environmental assessments of companies. In this new era, the view of environmental problems has undergone a shift from localized industrial pollution to a broader realization of their collective impact on global environmental problems.

The industrial pollution problems that afflicted industrial nations were recognized in the 1960s. In most industrial nations, strong campaigns were undertaken to identify the source of pollution and the damage caused. In many cases, the source and dangers could be traced to specific businesses and specific actions. Programs were swiftly implemented that satisfactorily dealt with the situation via "end-of-pipe" regulations that restricted emissions of air and water pollutants from factories. In the process, industrialized nations have increasingly shifted their emphasis from direct end-of-pipe regulations on industrial pollution to more market-oriented measures that encourage creative solutions to reducing environmental loads and costs (short-term and long-term).

End-of-pipe regulations required companies to comply with emission standards for gases, water, and noise emanating from production sites. Companies complied with the legal emission standards by installing treatment facilities for gas and water emissions. However, the controls were viewed as being problematic because they restricted operations and increased costs. As such, pollution controls did little to encourage corporate competition to resolve environmental problems on a grander scale.

However, in the 1980s with the recognition of the dire long-term impacts of global warming, acid rain, and the depletion of rain forests, environmental problems became global in scale. These newly recognized problems were

1

not derived by end-of-pipe industrial pollution as much as from environmental loads being generated in the daily activities of businesses and individuals that when assessed individually seem small but in the aggregate have a huge regional and even global impact.

This new, more market-oriented regulatory approach to environmental problems is aimed toward encouraging companies to find creative solutions to their global impact by reducing carbon dioxide and other greenhouse gas emissions, curtailing household and corporate waste generation while actively promoting recycling, and introducing the PRTR (Pollutant Release and Transfer Register) system for reporting and registering potentially harmful emissions and transport of waste that attacks the problem on a large scale.

As part of this effort, green procurement initiatives were enacted that promoted purchases of environmentally friendly products. Also basic laws were developed to promote a recycling-oriented society and regulate the tracking of emission audits. In its totality, these efforts aimed at curbing emissions on the regional and global scale. These laws form the legal framework for waste management and recycling.

Another part of this new regulatory approach has been a growing trend to emphasize environmental considerations in the previously untouched product and service markets. Through packaging and recycling, consumers began sorting discarded packages into recycling categories (e.g., glass bottles, PET bottles, steel and aluminum cans, and paper packs). This sorting allowed municipal waste collectors to collect these packages separately and business operators the opportunity to retrieve and re-use them. A similar approach emerged regarding the recycling of industrial waste. The approach allowed initial recycling costs to be borne by businesses through prudent pricing passed to a degree to consumers.

Through laws and reorganized handling procedures, the cost of recycling packages was reduced to the point where it no longer internalized the societal costs and, as a side benefit, companies have begun adopting a long-term effort toward converting to lighter, more simple packages and reusable materials. Appliance recycling efforts (e.g., air conditioners, televisions, refrigerators, and washers) allow consumers to return discarded appliances to retailers, who then deliver the products to their respective manufacturers. Consumers must pay the recycling costs, in this case at the time of return. The fee is to help suppress excessive consumer consumption.

As time has passed, it has become increasingly clear that companies that adopt the most efficient recycling method enhance their price competitiveness, whereas those who recycle inefficiently suffer. Herein is found a fundamental enhancement from the end-of-pipe pollution regulations in the 1970s. The latter focused on environmental refund costs to comply with emission standards. Under the market approach, companies are encouraged to invest in ways that will differentiate their products and enhance competitiveness.

Energy conservation efforts invoke even tougher market principles to encourage companies to take creative steps in reducing energy consumption (and hence, carbon dioxide emissions). Key features include the setting of energy conservation standards for designated machines (automobiles, home appliances, and office automation equipment) base. Competition is increased, driving each company's products to meet the standard set by the most energy-efficient product on the market. Companies that fail to improve product performance are subject to increasing market-driven as well as regulatory-driven forces. Those companies who cannot (or will not) compete are ultimately forced to withdraw their products from the market. By compelling companies to meet the highest energy-saving standards on the market, the global legal market affects not only the competitiveness of products but strikes at the heart of overall corporate survival.

Pollutant release and transfer laws emerged in the industrial community worldwide, requiring companies to measure and report emissions for specified chemical substances released into the air, water, and soil, as well as wastes transferred offsite for treatment and disposal. The impact of this tracking is that companies are increasingly performing risk management of chemical substances—the cost of which is often quite substantial—and when recognized, induce efforts to reduce, mitigate, or eliminate.

The best policy is to avoid using any harmful or hard-to-manage chemicals but often this is unrealistic. Regardless, a system should be in place to minimize the generation of harmful chemical waste via manufacturing methods and processes, raw material selection, and product compositions. This improvement will have significant implications not only in preventing direct contamination risk but in avoiding the risk of marketplace rejection or exclusion from customers and consequent competitive disadvantage.

Both the manufacturing and retail communities are increasingly recognizing the market power of green purchasing. *Green purchasing* refers to the activity of purchasing products and services with the smallest environmental impact, thereby encouraging companies to become more environmentally responsive. The approach is based on the idea that "green consumers" use their purchasing power in the consumer market to reward businesses that actively engage in environmental issues and to prompt reluctant businesses to do more.

However, whereas there is increasing evidence of consumers' awareness of green purchasing, it has been difficult to establish the degree to which green purchasing is having an impact at the consumer level. There is a surprisingly strong response to purchasing within corporate commitments that are trying to establish "green credentials." To that end, the green purchasing movement has been rapidly growing among businesses, organizations, and government offices, thereby effectively promoting green markets. There is hope that this will increasingly shift market demand toward green products, thereby leading to lower prices. More progress has occurred as cost savings are increasingly identified through these "green procurement" efforts. Large assembly and processing companies are increasingly procuring raw materials and components meeting their environmental specifications. Markets are clearly becoming more selective regarding environmental factors. In the future, even more emphasis is expected in business-to-business green transactions.

While the environmental laws impose significant cost burdens on companies, they, along with market factors, provide an impetus for companies to respond with creative solutions and business decisions to environmental problems (see Exhibit 1). To that end, under the energy environmental market-oriented approach, corporate survival will rest on the success of business strategies (environmental strategies) to enhance price competitiveness and business performance.

In the manufacturing sector, reducing environmental loads in the various stages of manufacturing, use, and disposal of products will require

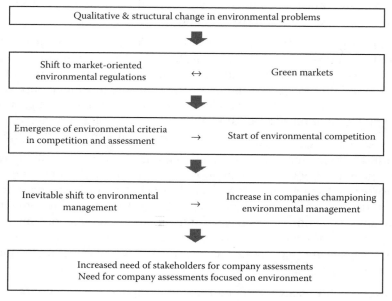

Exhibit 1. New environmental problems and the environmental competition of companies.

ongoing review of raw material and component selection and procurement, production processes, distribution channels, packaging, and volume and ease of dismantling and recycling wastes. This fosters a "design for the environment" (DfE) approach based on the product life cycle assessment (LCA) of environmental loads. In turn, life cycle costing (LCC) that results from these design efforts will impact price competitiveness. This adds up to companies that will clearly be differentiated by the strategies they adopt. Such re-engineering cannot be achieved with a "follow-the-leader," conformist mentality, such that the environmental re-engineering has spawned a new competitive condition.

Therefore, as green purchasing and procurement move ahead, the market forces are increasingly pressuring companies to develop more environmentally compatible products and services. In the past, companies pursued products based on price, performance, and quality standards. In the future, the additional competitive condition of the environment will force companies to also demonstrate "environmental rationality."

As a result, companies are reducing environmental loads across all activities and developing methods to measure and verify their green energy. Still, the objective assessment of corporate environmental performance is difficult both internally and externally. To that end, an effort is made here to lay out a framework and identify environmental management assessment and reporting standards. Thus, the environment has emerged simultaneously as a new assessment standard as well as a competitive condition.

Environmental considerations historically have little effect in determining a company's competitive position, nor have companies been systematically scrutinized with environmental management assessments. But the rules of competition are shifting to include environmental rationality. Companies are increasingly confronting a new era of environmental-based competition, where to an increasing degree business success is tied to competitiveness in environmental factors.

Until recently, Corporate Environmental Management has been largely reactive in nature, reluctantly complying with environmental controls while being monitored by regulatory and stakeholder groups. However, the tidal shift toward green markets is fostering a more proactive stance on environmental management that anticipates new regulations, market trends, and environmental competition factors.

Environmental management in the classic sense refers to the development and execution of environmental strategies to assure sustained corporate growth. The crux is the implementation of an environmental management system (EMS) based on seven principles. Of these, corporate commitment is the key; without it, no Corporate Environmental Management program can succeed. But how to measure it? And how does one measure the overall concept of Corporate Environmental Management? There are seven principles with which one can begin:

1. Commitment of top management
2. Development of an environmental plan and organizational structure
3. Company-wide participation of all employees
4. Efficient use of management resources
5. Sustained effort to reduce waste generation
6. Detection and minimization of environmental risks
7. Disclosure to stakeholders and investment community

A growing number of companies are establishing environmental management capabilities via environmental performance reporting, sustainable development reporting, and since the mid-1990s, ISO 14001 certification. Companies have been even releasing environmental audits and management assessments, but there is a bevy of parties interested in seeing such assessments formally developed.

Besides consumers, business customers, and suppliers, companies are surrounded by a variety of stakeholders—groups with an interest in a company's activities such as shareholders, banks, investors, local residents, and government agencies. As the severity of environmental problems has become clearer, stakeholders have increasingly demanded that companies be assessed on their environmental risks and merits that potentially impact their investment.

Just as interested are financial institutions, most notably when making loan or investment decisions. There is a strong movement to increasingly consider a company's environmental stance and environmental risks to a much higher degree than in the past, when environmental factors were excluded from consideration because of difficulties in determining their effect on a company's financial performance. But it is now understood that corporate survival has an environmental dimension, so whereas it is unlikely that poor environmental management will result in immediate business failure, there is little doubt that in the future competence in environmental management will become an increasingly important factor in price competitiveness, financing, and business performance. In short, corporate survival will rest in part on environmental factors, and poor environmental decisions and unacceptable environmental risk could prove damaging. Asset managers are interpreting more broadly the prudent man rule—minimize risk, maximize returns, and preserve assets—to encompass the environmental performance/risk issue.

Finally, international investment companies are now selling "eco-funds" (environmentally responsible stock investment trust funds). These funds are groundbreaking financial products in two respects: for confronting the issue of environmental responsibility and for investing selectively in

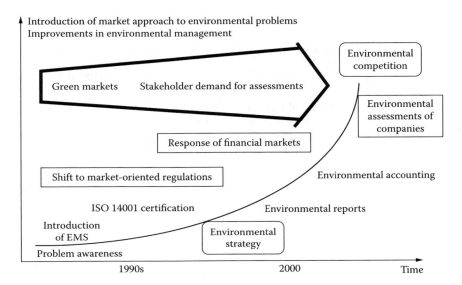

Introduction of market approach to environmental problems
Improvements in environmental management

Green markets   Stakeholder demand for assessments

Environmental competition

Environmental assessments of companies

Response of financial markets

Shift to market-oriented regulations

Environmental accounting

ISO 14001 certification   Environmental reports

Introduction of EMS

Environmental strategy

Problem awareness

1990s   2000   Time

Exhibit 2. Evolution and assessment of environmental management.

companies based on environmental criteria. Exhibit 2 lays out some of the forces generating the urgent need for Corporate Environmental Management assessment.

Reference

Kawamura, Masahiko. *NLI Research No. 145, 2000* (NLI Research Institute, 2000).

Chapter 2
Environmental Management Assessment

Given the uniqueness of environmental management issues and its position outside the mainstream of the company's day-to-day operating activities, it can be difficult for corporate management to assess the state of its program. One way to garner such independent assessment is to engage an environmental management assessment. An *environmental management assessment* provides a unique perspective that encompasses a senior management viewpoint along with a critical, technically astute, regulatory agency perspective. As such, the environmental management assessment can provide a critical assessment of the integrated fit of a company's environmental management program, an independent measurement of its effectiveness, and the timely identification of any potentially serious deficiency, as well as specific recommendations for programmatic improvements.

The central question that must be asked is, "What will be the role of the environmental health and safety (EHS) function?" Is the EHS function a mere commodity to achieve efficient compliance or is it a growth function based on a resurgence of societal pressures and regulation? Taking the latter point a step further, it can be argued that EHS has become a critical element of a successful 21st-century business strategy. Furthermore, as depicted in Exhibit 3 there are potential business consequences for the "efficient compliance" approach.

It is our view that the "best practice" EHS operating model is the "business partnership" model. This model calls for creating a strong EHS component in strategic planning and investigating beyond required compliance. In short, the company incorporates a design for environment philosophy into product development and process design. Under this approach, EHS issues are recognized as meaningful components of overall value propositions. There is also a concerted effort to not only measure EHS returns over time but to understand that, as noted in Exhibit 4, a severe lack of aggressive EHS compliance can result in even greater long-term costs.

9

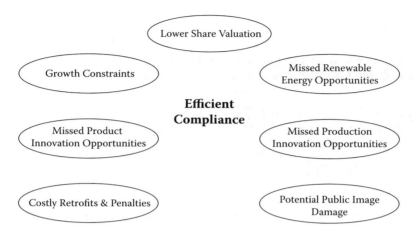

Exhibit 3. Business consequences of the "efficient compliance" approach.

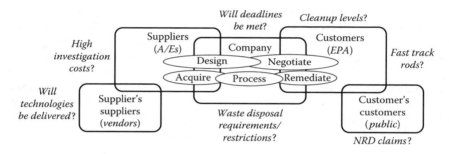

Exhibit 4. Few environmental restoration chains are effectively integrated.

The environmental management assessment analysis should employ tested methodologies for extracting key issues that may be adversely influencing the company's environmental management program. As discussed in the text and shown in Exhibit 4, the company's environmental management program can be strongly influenced by the degree to which the effort is integrated. It is our hope that the environmental management assessment analysis will enhance both the understanding and the implementation of such integration.

To be successful, the environmental management assessment should be an upper management exercise, preferably a board of directors-generated exercise. While significant day-to-day coordination and cooperation will be needed at both the operating and environmental management levels, it is essential that results be transmitted directly and freely to the highest corporate levels to ensure independence.

Recognize that CEOs and board members face significant responsibilities and liabilities with regard to environmental management performance. An independent environmental management assessment provides both a practical method for minimizing exposure to liabilities as well as a sound legal defense for making appropriate efforts to ensure successful environmental management performance and the assessment of exposures.

Also recognize that at first glance, some uncertainties may lie outside the control of environmental management. These uncertainties relate to unresolved issues, undeclared agendas, or responses by parties to the environmental management process. Effective environmental management seeks to identify, early-envelope, and convert these uncertainties into known factors that can be included in the overall management plan.

Besides dealing with uncertainty, the objective of the environmental management process should also be to increase the overall effectiveness by which organizational resources committed to environmental programs are used. In essence, environmental management programs must effectively integrate regulatory, technical, and management issues to provide well-rounded, cost-effective environmental management solutions.

The focus of the environmental management organization should not be limited to overall environmental management goal setting (i.e., the "bully pulpit" or sermonizing) but should include detailed technical planning responsibilities, regulatory documentation preparation, and establishing environmental cost estimation and training protocols to ensure the desired results are achieved. The process should also encompass the detailed preparation of critical environmental regulatory documents and corporate positions as well as technical support information that provides a strategic value to the company and the company's cost, financial, schedule, and regulatory objectives. Select activities should not be treated as individual events but as part of the total view to the environmental management problem identification and resolution process. As a result, the process generates a greater understanding of potential resource requirements and the impact of technical and regulatory hurdles to meeting environmental management goals.

Proactive Corporate Environmental Management must recognize that environmental risk is more than the assessment of individual company risks due to accidents, major incidents, unexpected regulatory cost impacts, or unexpected damage to brand reputation. True environmental risks are the bottom-line impacts on the stock of the company. To that end, the objective is to provide the investment community with environmental information that accurately reflects the company's environmental management performance and that can be translated in hard financial terms.

Chapter 3
Lines of Inquiry

The objective of Corporate Environmental Management is to increase the overall effectiveness by which organizational resources committed to environmental restoration are used. An independent *environmental management assessment* (EMA) evaluates the degree to which management programs effectively integrate regulatory, technical, and management issues to provide well-rounded, cost-effective environmental restoration solutions (Exhibit 5).

The assessment focuses on not only overall environmental management goal setting but also the adequacy of the detailed technical planning, regulatory documentation, and cost estimation vis-à-vis the results desired and achieved. Undertaken from a senior management perspective, it also encompasses detailed reviews of critical environmental regulatory documents and technical information from the standpoint of their strategic value given cost, schedule, regulatory, and overall corporate objectives.

EMAs must take a system-based, "big picture" approach, not auditing select activities as individual events but as a part of a total view to environmental management problem identification and resolution. As a result, the assessment generates a greater understanding of potential resource requirements and the impact of technical/regulatory hurdles ("showstoppers") on meeting environmental management goals.

Three major lines of inquiry are pursued during the EMA process: they are an "internal survey," an "external scan," and a comparative analysis (or "benchmarking") survey. As shown in Exhibit 6, the three-point environmental management audit provides a company with three different perspectives on environmental management performance. Thus, the company receives feedback that reflects not only performance within the organization but also reveals the organization's position relative to the community as a whole and reflects the expectations and realities of the company's public image.

Taken together, these three lines of inquiry provide a comprehensive basis for evaluating a company's environmental management performance. The first two lines—"internal survey" and "external survey"—are a basis for establishing a company's specific environmental management profile. In turn, "benchmarking" is a comparative exercise evaluating the

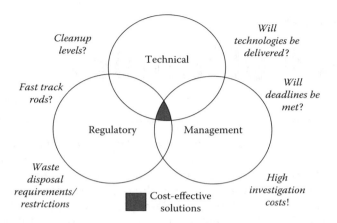

Exhibit 5. Integrated environmental management programs.

Exhibit 6. Components of an SC&A environmental management assessment.

company's profile versus a selected set of other organizations and their relevant environmental management case histories. The particulars of each line of inquiry are discussed further.

The first component is an *internal survey* of key environmental personnel and various key interface personnel throughout the company's operating units. The internal assessment confirms the effectiveness of the company's environmental management program, both from a compliance issue with management and from an internal culture development standpoint. It addresses the degree to which an organization's environmental management is based on well-defined policy and programs or whether it relies heavily on "informal networks" and "individual initiative."

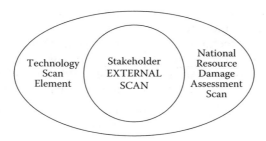

Exhibit 7. Elements of an external survey.

The internal survey involves a review of available internal documentation and internal procedures that define a company's environmental management program. The scope of the assessment includes:

- A review of policies, guidelines, and procedures relevant to the control of technical quality;
- A review of cost and schedule estimating processes;
- An independent cost and schedule review of a statistical sample of projects across the company's environmental management universe;
- A review of a statistical sample of monitoring data for compliance with data quality objectives and cost-effective regulatory strategy;
- An evaluation of the outside environmental support contracts for their ability to control contractor activities from a technical, cost, and schedule standpoint;
- An evaluation of the control processes for activities funded by direct and indirect charges;
- An evaluation of the technical/regulatory decision-making process and documents prepared;
- An identification and assessment of technical/regulatory impacts on cost and schedule via value engineering and cost benefit studies; and
- The identification of contingency management and enhanced cost control opportunities.

In addition to selected file and document reviews, on-site visits are conducted at each of the company's major environmental restoration and plant sites.

As shown in Exhibit 7, the external survey has three potential components. The *external scan* component is a confidential canvassing of key stockholders (governmental, regulatory, public advocacy) for their impressions of the company's environmental management and performance. As a supplement, the external survey may also involve an independent technology survey scan and an independent natural resource damage assessment scan that provide the company with independent feedback of key issues that define the ultimate success of its environmental restoration program. The independent technology scan element assesses whether existing company environmental technologies are current or in danger of being outdated given energy technologies or regulatory criteria. Last, a confidential natural resource damage assessment scan may be added to evaluate whether or not the company is vulnerable to a claim in the making.

The last line of inquiry is benchmarking. The *benchmarking* efforts entail a comparative analysis of environmental restoration experiences, policies, practices, and programs between the company and selected like enterprises—giving particular emphasis to "regional organizations" whose site characteristics match the company and other national organizations of similar industry characteristics that are perceived to be "industry leaders." Principal emphasis is placed on determining "best practices" so the company not only gauges its relative degree of accomplishment but also can establish future performance goals based on an awareness of the "best of the best" in environmental restoration management approaches. To complete the benchmarking exercise, information may be secured (through on-site visits) with each of the selected benchmarking companies, vis-à-vis independent consultant support. Under this approach, each company agrees to participate on the basis that its own individual data and information would be kept confidential by the independent consultant.

Chapter 4
Assessment Model and Analytical Framework

As a useful way of measuring the company's environmental achievement and improvement opportunities, reliance is placed on a "stages of excellence" model developed for assessing the performance of numerous business functions (Exhibit 8). The *Stages of Excellence Model* is a consistent measure that the organization can apply to its environmental management relative to other operating units' performance.

Assessment Model

The key premises of this model are that:

- Major activities (and their component elements) progress in maturation through different steps or "stages," from their formation at an experimental level (Stage I) to the achievement of a truly outstanding leadership position at Stage IV.
- In most cases, the transition from one stage to the next stems from a breakthrough or "trigger point."
- Component elements of a function, such as environmental management, span several stages at any particular time.
- Advantage can be gained by determining the most advanced elements of functional excellence (in stages) and the least advanced, and then deliberately focusing on the latter on a "management-by-exception" basis to strengthen overall corporate effectiveness.

Environmental management is a unique corporate management issue in the sense that it is typically "outside the mainstream" of the company's profit-generating operating activities and is subject to heavier reliance on outside contractors. However, it is still critical to assess the performance of these programmatic activities in keeping with the company's overall comprehensive mission goals. Furthermore, it is critical to assess the benefits (or lack thereof) received by contractor support in driving the company's environmental program up the "stages of excellence" curve. For example,

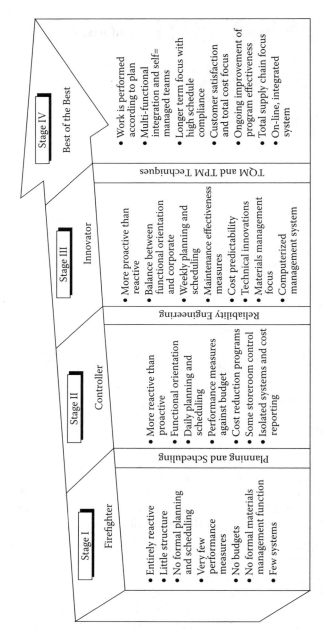

Exhibit 8. Environmental management assessments: Stages of Excellence.

except in isolated cases companies typically will not look to internal resources for achieving technology performance breakthroughs. However, it is fair and reasonable to evaluate contractor support in that regard.

As applied to environmental management efforts, it is clear that at the *Stage I* level the organization is in an experimental stage and, for the most part, in a reactive mode, subject to unexpected surprises with key regulatory, management, and operational systems being undefined or, to some extent, unrecognized. Typically, at Stage I the performance of the various elements of environmental management varies in an inconsistent fashion.

At a *Stage II* level, some consistency in environmental management is being achieved. However, the environmental management is very control-oriented with very little company-wide cultural development and innovation.

At the *Stage III* level, performance breakthroughs are occurring from technical, regulatory, and management standpoints. Problem identification is occurring in a proactive manner accompanied by more timely identification of solutions. Technology alternatives are being identified that have a positive impact on cost, safety, or schedules. Regulatory relations are well established. Management systems are in place that provide a greater level of predictability regarding technical performance, cost control, schedule control, and consistency in management direction.

At a *Stage IV* level, significant benefits are arising from well-integrated programs. Relations with the regulatory community reflect a degree of flexibility based on trust and proven technical performance. Management systems have been developed that are resulting in significant productivity gains, both due to the optimal arrangement of programmatic activities, enhanced "lessons learned" communication within (and outside) the organization, and significant movement up the learning curve for technologies in place.

Analytical Framework

As a useful way of measuring the company's environmental achievement and improvement opportunities, reliance is placed on the environmental management system framework presented in Exhibit 9. This framework was developed to provide a consistent, across-the-board method for assessing the proficiency with which environmental management activities are managed. Under this framework, environmental management is gauged in terms of 14 principal analytical criteria divided into four broad categories—Corporate Commitment, Functional Implementation, Communication, and Measurement and Control Systems. Thus, findings in each of the foregoing inquiry areas (internal survey, external survey, and benchmarking) are concentrated on these subject areas and reflected within the Stage of Excellence Model. Recommendations and implementation priorities can

---

CORPORATE COMMITMENT CRITERIA

- **Senior Management Commitment,** which addresses the extent of top-level leadership of, and support for, corporate environmental activities, and the provision of adequate resources for them.

- **Corporate Environmental Policy,** which addresses the formal delineation of corporate environmental standards and expectations, and the articulation of guidelines and principles by which a company plans to achieve its vision.

- **Strategic Environmental Plan,** which translates policy into a concrete framework for implementation; includes assessment of resources needed to achieve expected performance; documents goals, milestones, performance expectation, and ways to measure achievement or lack thereof.

COMMUNICATION CRITERIA

- **External Communication/Public Relations,** which addresses involvement with external constituencies such as environmental regulators, elected officials, customers, and the public and includes public outreach programs, lobbying efforts, and intervention, where appropriate, in regulatory proceedings.

- **Internal Communications,** which includes (a) vertical communication to employees about company environmental goals and policies (and provides for employee feedback) and (b) horizontal communication between the environmental functions and other areas of the company.

- **Issue Management,** which addresses ways in which major environmental issues are identified, addressed, and resolved. This extends to "early warning systems" for new issues and ways in which accountability for issues management is assigned.

Exhibit 9. Principal analytical criteria.

---

then be structured around these 14 criteria to drive the environmental management performance up the stages of excellence curve.

Program vs. Project Management

A central question to ask is whether the company is better served considering the environmental restoration activity as a "program" versus a "project." Inherent within the title "program" is a greater emphasis on development of internal resources for managing an ongoing core company mission via staff development and equipment acquisition.

In many cases, environmental management is a unique mission outside the mainstream and outside the scope of most corporate activities. Thus, in some instances it may even be better for companies to consider the environmental management mission as a "project" management exercise where technical resources are largely contractor-supplied and the company's environmental management is focused on "project management." Exhibit 10 provides a schematic way to assess the issue. The company is best focused on "doing" those activities where it has proprietary capability and value-added support. Whereas those functions may be essential and even proprietary, they should also be continually reevaluated and with time, if appropriate, moved to the "buy" category.

FUNCTIONAL IMPLEMENTATION CRITERIA

- **Organization and Staffing**, which addresses appropriate structural positioning and channels of communication for the environmental functions and the qualitative and quantitative adequacy of environmental staff resources at all levels and in all areas of the enterprise.

- **Internal Integration,** which addresses the extent to which environmental goals, programs, and priorities are embraced and pursued by all elements of the enterprise and reflected in planning and decision-making throughout (not just within) the environmental functions. This addresses linkages between environmental and closely related health and safety programs.

- **Operational Systems,** which address the scope and effectiveness of day-to-day environmental operations, provisions for dealing with unusual events and emergencies, and the operation and maintenance (including preventive maintenance) of major environmental facilities and/or equipment.

- **Risk Assessment,** which addresses the thoroughness, completeness, objectivity, and candor with which all types of environmental risks are identified and assessed throughout the company, and programs developed to mitigate such risks.

- **Waste Minimization,** which addresses programs to develop and implement activities dealing with waste reduction, pollution prevention, and recycling.

MEASUREMENT AND CONTROL SYSTEMS CRITERIA

- **Audit Program,** which addresses the comprehensiveness and effectiveness of environmental auditing activities, including the setting of standards (e.g., compliance of "beyond compliance"), definition of protocols, involvement of line management, and implementation of results.

- **Environmental MIS,** which addresses the availability of easily accessible on-line systems to track environmental performance, activities, and issues.

- **Financial Tracking,** which addresses the availability of accounting systems for environmental expenditures of all types and a system to track overall expenditure trends, including capital and operating and maintenance (O&M) charges.

Exhibit 9. (continued) Principal analytical criteria.

	Program Management Emphasis	Project Management Emphasis	
	Industry demands a heavy investment in environmental health and safety	Industry does not typically demand a heavy investment in environmental health and safety	
Unique Problem: value-added support required	Exceed Standards Develop best capability internally	Develop Access to Best capability within a cost/benefit	• Significant Mission Scope (Size)
Standard Problems: basic support required	Meet standards	Develop Access to Capability that ensures compliance	• Limited Mission Scope (Size)

Exhibit 10. Program versus project management analysis.

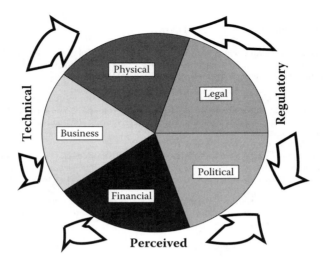

Exhibit 11. Interwoven risk perspectives.

Where heavy emphasis is placed on using outside subcontractor resources, a company's environmental management philosophy should be structured to maximize the potential for sharing cost/schedule risk and management risk with subcontractors under well-designed incentive programs.

Environmental Risk Management

Environmental risk can be defined as an event or condition that can result in corporate reputation damage or material financial loss or that prevents the company from achieving its business objectives. Environmental risk management may encompass three types of risks:

- The technical risk (established by site service and regulatory/ agency personnel);
- The perceived risk (outrage) by the public; and
- The regulatory risk relative to past, current, and future standards and position.

However, as shown in Exhibit 11 within these three outlying risk perspectives are interlocking risk sub-elements (physical, business, financial, political, and legal) that are derivative of core business functions and constraints.

Technical risk is measured in terms of outcomes, magnitudes of the outcomes, and probability of occurrence. It also identifies and gauges the impact of the known and unknown elements that factor into the risk assessment. At heart, technical risk management is an engineering/ science exercise. Typically, industry managers think of technical risk when considering risk management.

• Involuntary	• No Control
• Unfamiliar	• Media Attention
• Distrust	• Unfair
• Irreversible	• Future Generation Impacts

Exhibit 12. Perceived risk reflects public attitude.

Perceived risk reflects the public attitude. It is critical to understand that the public's ranking of perceived risk may not be highly correlated with actual technical risk. Exhibit 12 presents a suite of issues and concerns that could fall within the perceived risk assessment. However, each site is unique and its perceived risk will reflect its unique conditions. Perceived risk is perhaps the most difficult of all risks to anticipate and deal with efficiently. For example, it can be manipulated by activist groups interested in controlling or defeating the site restoration project. Perceived risk can be skillfully modified and presented to arouse the sympathy of the media, politicians, and the regulatory community. However, the regulatory group is generally resistant to this type of pressure and if the restoration approach is not well-planned and presented, it may suffer defeat in the public forum. Regulators can only evaluate and make the case presented to them.

Regulatory risks cover compliance with federal and state standards as well as compliance with corporate (and in the cases of Department of Energy and Department of Defense, governmental agency) policies. Regulatory risk analysis should be approached with a proactive attitude. This means assessing the potential for retroactive application of future standards (e.g., Superfund) and the latter's implication for the company. It also means the development of future internal standards that reflect the public's risk perceptions as well as currently established technical and regulatory risks. Again, it should be recognized that future regulatory risks typically are politically driven; they do not necessarily reflect technical risk. However, regulatory risks are the ones that ultimately dictate cost to a far greater degree than technical risk and, as such, require as much emphasis, if not more.

Overall environmental management risk is an additive of these components. The three types of risks must be integrated in an environmental management risk review process that includes identifying and assessing risks, prioritizing risks, identifying alternatives, analyzing the alternatives, and selecting and implementing a strategy. Identifying and assessing risks is a process of developing an understanding of the business consequences of each risk component. Prioritizing risks is a process of combining the technical and regulatory risks with the perceived risks. Identifying risk

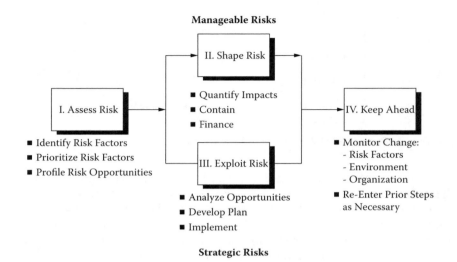

Exhibit 13. Environmental management risk review process.

management alternatives involves not only assessing the technical and regulatory options but also the outreach/communication options available. In some cases, alternative strategies may require a combination of technical, outreach, and other elements. Exhibit 13 provides an illustration of an environmental management risk review process.

The alternatives should then be analyzed for their cost/benefit and also from the standpoint of the uncertain versus certain composition of the cost/benefits. Implicit within this analysis is the crucial question of feasibility—technically, regulatory, and public acceptability—all of which impact cost feasibility. Central to the analysis are the questions, "How much are we willing to pay for the 'uncertain benefits' of a strategy? Relative to the cost, how much uncertainty are we willing to live with for specific options?"

For example, the question might be asked, "How is this risk of failure in the remediation waste management concept to be measured?" Central to the measurement of the potential failure of a program is the *Technology Risk Assessment* (TRA). This can be defined in a systems approach to identify and evaluate the risks associated with a given remediation or waste management technology. The TRA must consider risks associated with technology failure, indirect consequences, and primary and secondary risks of accidents and malfunctions. The TRA does not duplicate or replace human health or ecological risk assessments but supplements them with the boundary condition of realistic technology expectations.

In short, environmental risk management requires integrating risk management analysis with the key business management processes of the

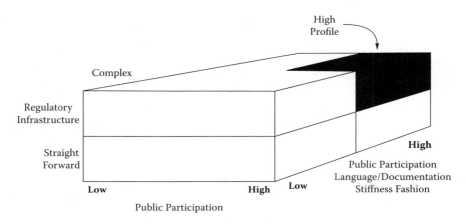

Exhibit 14. Environmental management regulatory model.

organization and actively managing technical, perceived, and regulatory risks. Environmental risk management is part of the company or site's job and must be integrated into the overall environmental management processes. It also includes addressing a public participation element that provides the company with independent feedback of key issues that define the ultimate success of its environmental program. This is critical to establishing a regulatory model.

As shown in Exhibit 14, it is critical to establish the company's environmental management regulatory model that applies to the critical company environmental program efforts. The regulatory climate assesses the public perception relative to a company's overall role as an ombudsman for the public's environmental trust as well as the regulatory infrastructure. The latter may be well-established. In some respects, that can be beneficial because regulatory requirements may then be well-defined, in contrast to an undefined regulatory environment that leaves critical environmental requirements undefined from a political standpoint.

Alternatively, we are all familiar with the debilitating effects of needless, complex, and confusing regulatory requirements. Thus, we add in a third element: the language/documentation stiffness factor. This reflects the complexity of the documentation requirements that may be required in the case of international environmental programs, the complexity of language translation requirements and, in domestic cases, often the challenge of translating from "techno-speak" and "legal-speak" into common, understandable language.

Chapter 5
Internal Survey

The *internal survey* involves evaluating the two distinct missions that most company environmental managements face—namely, providing support for:

1. The corporate headquarters' environmental activities, and
2. The company's plant and operating environmental support needs.

While interrelated, the two missions have distinct and separate challenges. Corporate headquarters support activity is a staff function by nature whereas supporting plant/operational needs, if it is to be successful, must involve some level of field operational buy-in to achieve effective integration.

Corporate Headquarters

The corporate headquarters mission involves coordination of the company's strategic environmental decisions. Activities such as tracking environmental legislation, evaluating potential economic impacts, and coordination with lobbying efforts naturally fall within this realm. Likewise, the establishment of corporate environmental goals and coordinating and maintaining the corporation's key permitting and regulatory compliance framework falls within the realm of a corporate environmental staff mission. However, the hard realities of implementation at the plant/operations level bring up a series of questions and differences in environmental management philosophies.

How much control does the corporate environmental management need over plant environmental support personnel? As shown in Exhibit 15, in some instances tight operational control over plant environmental activities may be warranted when serious environmental issues are a necessary part of operational activity or environmental restraints have significant impact on plant production capabilities. In other cases, a corporate environmental staff may function in a "bully pulpit" fashion—setting the framework for compliance (e.g., SOPs, permits) or providing some level of internal audit but leaving day-to-day implementation responsibilities to the operational management. There is no one correct approach. Selection of an environmental management style depends on company culture, industry specifics, and organizational strengths and weaknesses.

	Minimal Financial Risk	Significant Financial Risk
Complex Environmental Mission	Bully Pulpit	Driving Dad
Straightfoward Environmental Mission	Friendly Uncle	Encouraging Dad

Exhibit 15. Positioning the environmental management role.

⇐	Corporate Line	⇒
	Major issue management of regulatory negotiation	*Internal integration*
⇓		⇓
		Plant walk-throughs

The internal survey of environmental management should emphasize breadth with selected depth as dictated by management/board/public concerns.

Exhibit 16. Internal survey, examples of breadth versus depth.

Using the four assessment categories and their 14-key criteria for success (presented previously) as the framework for the internal survey, the environmental management assessment team engages in:

- Discussion and review of environmental policies and procedures;
- Review of environmental records;
- Plant walk-throughs;
- Discussions with environmental personnel; and
- Discussions with corporate and line personnel.

The internal survey does not necessarily need to be an audit per se. In fact, there is a strong argument for keeping it oriented more toward breadth than depth, as shown in Exhibit 16. However, the survey should bore deeply on a selective statistical basis and focus on areas of particular concern to the board of directors.

Operations

Elements of an effective risk management program at the plant level include evidence of an ongoing risk characterization and reduction program; episodic risk management systems reviews that include plant operating personnel; evidence of process (operational) safety standards and their implementation; evidence that EHS standards are being applied uniformly across facilities and (if applicable) being applied worldwide; evidence of a proactive emergency response, community communications, and involvement program; a product stewardship culture in place; and evidence of effective change management.

The area for effective environmental change management is broad and goes beyond pipes, pumps, control systems, and information systems. For example, it can include changes in information systems such as integrated software systems covering all business activity, including EHS data. It is critical to ensure that information management systems are developed that readily flag unacceptable EHS risks to concerned management.

There should be clear evidence that environmental risk management at the operations level is being driven by responsible management with the support of full-time EHS staffing experts. The said EHS subject matter experts should be tied operationally into project activities. At the very least, there should be a full-compliance philosophy with well-kept logs of potential concerns for tracking and resolution.

It should be clear that the work process owner (plant operations management) is responsible for mitigating risks. Environmental risk mitigation plans should be in place prior to considering system change and supported by a cascaded endorsement process including EHS senior management approval of acceptable risks.

Chapter 6
Corporate Commitment

Public trust is fragile! Corporate commitment is essential to maintaining public trust. The key criteria for the corporate commitment are:

- Senior management commitment
- Corporate environmental policy
- Strategic environmental planning

These are the three legs that must equally carry the weight of the public's trust in the company's environmental management and stewardship.

The overall focus of the corporate commitment category is to answer the question, "Where does management want to be from an environmental management perspective, and are the goals being attained?"

Senior Management Commitment

Senior management commitment addresses the extent of top-level leadership's support for corporate environmental activities, as expressed through the provision of adequate resources. It also assesses the overall impression among the firm's employees and stakeholder community regarding senior management's leadership and commitment to corporate environmental activities. Typical examples are a personal involvement to help resolve conflicts that occur with regulators, or proactively representing the company in industry forums and initiatives. This also includes assessing whether the senior environmentally trained management person functions at too low a level to adequately voice environmental issues to top-level leadership in a timely fashion. This point may be exacerbated if recent CEO turnover or management turnover has occurred. Because of these concerns, there may be a need for a conscious program to promote the CEO to a more visible environmental leadership profile.

Executive management should reassess their commitment to environmental values with an eye toward promoting and expanding the company's environmental capabilities and accomplishments and assessing whether environmental performance is a differentiating factor in their overall industry performance. Often, it is not sufficiently recognized that a company's past, current, and future environmental performance is a potential asset that can be "taken to the bank" for both maintaining and expanding its market area. Environmental performance can profoundly impact the company's ability to successfully negotiate with state, federal, and international governments for needed infrastructure development and to support needed expansion.

The following are some examples of ways in which companies can emphasize their senior management commitment in a company.

The CEO and COO could conduct meetings with employee groups several times a year where environmental management is a consistent theme. The company's board of directors could also adopt one of the established environmental principles, the business charters for sustaining development in an environmentally responsible manner.

A critical step toward developing a stronger board/senior management commitment and influence on environmental issues is to establish an environmental subcommittee on the board of directors authorized to enact environmental management principles and garnering independent talk-back on company environmental performance.

Another step toward enhancing senior management commitment and the board environmental subcommittee's roles is by having a subcommittee board member "observe" on an annual basis an environmental audit and report observations back to the board. This will also enhance internal communication of the company's environmental commitment.

A good example of senior management commitment is the establishment, composition, and range of an environmental risk oversight committee. Such a committee can be used as a forum to promote risk awareness throughout the company. It also can provide a basis for independent assessment of risks on an integrated basis through a cross-functional perspective. It can be a platform for independent review of risk control and mitigation procedures and provide guidance and support to operating management to implement risk control initiatives.

To be effective, the committee should not be stacked solely with environmental expertise but should reach out to a broad spectrum of the company's management as circumstances warrant, such as shown below:

CFO	Treasurer
Operations	Information Technology
Internal Control	Human Resources
EHS	Quality Assurance
Legal	Marketing

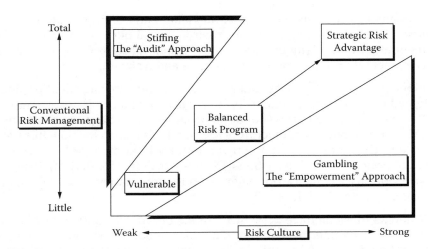

Exhibit 17. Risk management "audit" versus "risk" management cultures.

The goal should be that no "surprises" are found and that environmental risk management is built into the company's culture. The committee can be used to identify and to strengthen contingency plans where they are weak in those areas. It can also be used as a process to align management perceptions through more thorough communication.

However, a word of caution about controls: There is a difference between focusing exclusively on risk management using an "audit" approach versus developing a risk culture "improvement" approach. A more integrated approach often proves to be the best for the more complex companies (Exhibit 17).

Corporate Environmental Policy

The corporate environmental policy addresses the formal delineation of corporate environmental standards and expectations and the articulation of guidelines and principles by which a company plans to achieve its vision. If the company does have a formal environmental policy, is it highly visible to the informed public and company personnel? Is there at least an informal understanding among those surveyed that the company does have a goal and policy vision regarding environmental standards and expectations and strives to meet them? What is the company's environmental goal? Does it aspire to a world-class leadership position in environmental activities or does it wish to present a simple conscious commitment to environmental stewardship for above-minimal compliance?

In the past five years, companies in every industry have stepped forward to successfully establish environmental positions. This reflects a need for a company to revisit its environmental vision and policy, not so much to redefine itself but to distinguish its environmental performance from that of its competitors. Many companies find it helpful to draw upon outside environmental practice codes for guidance and a starting point.

In the past decade, private codes of environmental management practice emerged as a major force in corporate environmental programs. Examples of these codes include the Global Environmental Management Initiative (GEMI), the Chemical Manufacturers Association's (CMA) Responsible Care Program, the Coalition for Environmentally Responsible Economies' (CERES) principles, the International Chamber of Commerce's (ICC) Business Charter for Sustainable Development, and the international environmental management standard, ISO 14000.

Whereas these all have their own unique perspectives, they have common features. First, each requires companies to adopt environmental management systems and to audit their progress toward the environmental goals. Second, to varying degrees each calls upon firms to involve outside groups. Third, the goal of the private codes is to induce management to adopt more responsible forms of environmental behaviors. However, none includes specific environmental performance standards that firms must meet, and only ISO 14000 requires third-party verification of firms' environmental systems, and this is more of a verification of program elements being in place versus technical performance.

However, private codes can provide the impetus for the "end of pipe" performance. Private codes can span the types of changes in corporate policy, organization, and strategy that will lead to environmentally sustainable industrial practices. Typically, regulatory-driven responses involve adopting pollution controls but often leave products and manufacturing processes virtually unchanged. These types of prevention strategies generally must be tailored to the particular circumstances of the firm and arguably may be better addressed via private codes. Last, private codes strengthen corporate legitimacy and provide a venue for forward movement separate from the sometimes adversarial relationships between regulators and companies.

Private codes allow companies to demonstrate corporate knowledge about and commitment to environmental improvement with the focus more on learning and directly beneficial action versus defensive actions and litigation. In summary, private codes foster long-term changes in the ways firms think about the environment and how they integrate environmental aims with other business objectives.

Responsible Care

The Responsible Care initiative was originally recommended to the Chemical Manufacturers Association (CMA) by its Public Perception Committee, whose mandate was to recommend some industry initiatives that could improve the legislative, regulatory, market, and public interest climate for the industry. The program was created in the wake of the Union Carbide accidents at Bhopal, India, in 1984 and Institute, West Virginia, in 1985, when public distrust of the chemical industry was strong.

The Responsible Care program has several components: a set of "guiding principles," six "codes," a public advisory panel, and executive leadership groups. The codes address community awareness and emergency response, chemical distribution, pollution prevention, process safety, employee health and safety, and product stewardship. Responsible Care-type initiatives exist in approximately 30 countries in addition to the United States and Canada.

The CERES Principles

The driving force behind the Coalition for Environmentally Responsible Economies (CERES) is to foster socially responsible environmental investment. CERES's primary goal is to institutionalize the capability for generating corporate environmental management data that could be used by investors in their decision-making. It seeks "consistent, comparable, and widely disseminated" data that would allow investors to analyze environmental performance in the same way they analyze corporate financial performance.

CERES drew upon a coalition of investors, environmental advocacy groups, and labor unions to develop a common set of environmental principles. The CERES principle covers:

- Protection of the biosphere
- Sustainable use of natural resources
- Reduction and disposal of wastes
- Energy conservation
- Safe products and services
- Environmental restoration
- Informing the public
- Management commitment
- Audits and reports

The tenth and final principle was considered most important, and it stated that companies must annually complete and make public a "CERES report" containing detailed information on corporate environmental practices.

GEMI and the ICC Charter

The Global Environmental Management Initiative (GEMI) was formally announced in April 1990, though it had its start months earlier when a group of corporate environmental managers from large firms in the chemical, electronics, consumer products, and pharmaceutical industries began to meet regularly to discuss environmental management issues. The focus was to create a forum to share strategies, to stimulate critical thinking, and to strengthen dialogue between themselves and the interested public. It has grown to over two dozen members, representing large companies from a diverse group of industries.

GEMI members worked closely with the International Chamber of Commerce (ICC) to draft the Business Charter for Sustainable Development, which contains 16 principles tailored to large multinational corporations. GEMI's participation ensured that the charter received support from U.S. industries. One of GEMI's initial goals was to bring the charter to life by developing a database to track implementation efforts and an environmental self-assessment program to guide companies in this process. However, unlike the other code organizations discussed here GEMI has not required its members to adopt or implement the ICC charter. During its first year, GEMI members brought together the concepts of total quality management and environmental management, coining the term *total quality environmental management* (TQEM).

United Nations Environment Programmes' Financial Institutions Initiative on the Environment

United Nations Environment Programmes' (UNEP) Financial Institutions Initiative (FII) was founded in 1992 with the purpose of engaging the world financial institutions on the subject of sustainable development. Currently, there are almost 200 signatories worldwide. Signatories typically use the principles within the initiative's statement as a framework for identifying and managing risks in their lending and underwriting practices. This reflects a shift in worldwide development settings where financial institutions are increasingly seen as de facto regulators and tribunals for venting international environmental disputes, and it also reflects an increasing "green" sentiment in the public and private investment community.

Environmental Banking Association

The Environmental Banking Association (EBA) is a decade-old, U.S.-based forum for financial institutions to address environmental issues. It operates in a collaborative fashion with the UNEP-FII and focuses on proactively addressing environmental risks and their impact on the bottom line of financial institutions.

World Business Council for Sustainable Development

The World Business Council for Sustainable Development (WBCSD) is a coalition of over 160 international companies. It reflects a commitment to sustainable development as measured by economic growth, ecological balance, and social programs. The emphasis is on eco-efficiency and corporate social responsibility. The member-led organization is governed by a council comprised of member company CEOs.

ISO 14000

The International Organization for Standardization (ISO) was formed in 1946 and is headquartered in Geneva, Switzerland. Its purpose is to facilitate standardization as a means of promoting international trade. Whereas its membership consists of the standards organizations of its 100 member nations, ISO has been an "industry-driven" organization.

ISO standards are documented agreements of technical specifications that companies use as guidelines to ensure that materials and products fit their purpose. In the early 1990s, ISO came under pressure to develop an environmental management standard. ISO's Strategic Advisory Group on the Environment (SAGE) was set up in 1991 to consider the appropriateness of an international environmental management standard. SAGE's findings indicated that the environmental management standard would promote a common approach to environmental management, much as the ISO 9000 series had for quality management. It was also found that an intentional environmental management standard would enhance a firm's ability to attain and measure improvements in environmental performance and facilitate trade and remove trade barriers.

As ISO set up its environmental effort, many of the world's major manufacturing countries were in the process of developing environmental management standards of their own. Some 400 representatives of U.S. industries—including ones from the chemical, petroleum, electronics, and consulting sectors—have participated actively in the development of ISO 14000.

ISO 14000 keys on distinguishing three types of *environmental performance indicators* (EPI):

- *Operational indicators* that measure direct potential stresses on the environment (e.g., burning fossil fuels;
- *Management indicators* that measure efforts to reduce or mitigate environmental effects (e.g., company spending on environmental training programs); and
- *Environment condition indicators* that measure environmental quality (e.g., ambient air pollution concentrations).

Of the three, operational performance indicators may be the most germane to environmental management performance. It is the most direct link between the company's individual internal practices and the external environment. Some have called it the company's "ecological footprint" that defines the company's role in creating environmental problems and generating solutions.

Increasingly, four common set operational indicators are being recognized as keys to measuring the pollution prevention and resource efficiency of products, processes, and services. They are:

- Materials use
- Energy consumption
- Non-product output (i.e., waste)
- Pollutant release directly to air, water, and land

These four EPIs draw from the ecological rucksack of Germany as well as the U.S.'s Toxic Release Inventory (TRI) approach.

Like that of ISO 9000, ISO 14001 framework encourages firms to hire third-party contractors to certify that their management systems are in accord with the standard. ISO 14001 is explicit in its requirement that companies identify the "environmental aspects" of their activities, products, and services that they "can control" and over which they can "have an influence."

- Establishing environmental goals and targets: ISO 14001 calls upon firms to "establish and maintain documented environmental objectives and targets."
- Measurement systems: ISO 14001 requires companies to maintain procedures to measure "on a regular basis" the "key characteristics" of their activities that have a significant environmental impact.
- Employee training: Employee training features prominently in ISO 14001. ISO 14001 calls on companies to self-audit "periodically" or "regularly" to ensure that they are in conformity with "planned arrangements."
- Rewards and penalties for worker performance: ISO 14001 is explicit in this regard, indicating that companies must make employees aware of "the potential consequences of departure from specific operating procedures."
- Third-party verification: Third-party verification is a requirement for ISO 14001 registration.

With respect to assessing releases, establishing measurement systems, and setting goals, ISO 14001 includes specific requirements. A distinguishing feature of ISO is its requirement for third-party verification to obtain registration. An ISO 14000 subcommittee is developing general guidelines for conducting and reporting life-cycle assessment studies in a "responsible and consistent manner," but ISO 14031 does not call upon firms to use life-cycle approaches.

Whereas we are not necessarily advocating the certification of ISO 14031, "Environmental Management Systems—General Guidelines on Principles, Systems, and Supporting Techniques," we have found that a company can often refine its environmental policy so that it fully conforms to ISO 14001 requirements without making any substantial or controversial changes. In turn, ISO 14001 provides a framework for defining a company's program and environmental policy. Gearing the company's environmental policy to be in conformance with ISO 14031 can be useful, not only from an improved regulatory status but also providing significant public relations value.

Issues required under ISO 14031 that should be addressed in a company's environmental policy include:

- The organization's mission, vision, core values, and beliefs;
- Prevention of pollution;
- Guiding principles;
- Coordination with other organizational policies (e.g., health and safety); and
- Specific local or regional conditions.

In some companies, senior executives add their signatures at the end of their corporate environmental policies, illustrating a strong senior-level commitment to the policy. In particular, the CEO's name and signature beneath the updated policy demonstrates in no uncertain terms senior leadership's commitment to the environmental policy and to the environmental program generally.

An environmental policy in conformance with ISO 14001 considers requirements of and communication with interested parties. Oftentimes, a company's policy may address this issue to a limited extent in a few paragraphs but never explicitly covers communication. It is recommended that a statement be added to the policy as follows:

"We will maintain open communication on environmental issues with regulatory agencies, environmental groups, customers, and employees."

ISO 14001 also requires conforming environmental policies to include a commitment to continual improvement and pollution prevention. An example statement is as follows:

"We will continue to improve our environmental programs and environmental performance."

Subject to review by the general counsel, the company should consider adding the following phrase:

"...it will be in compliance with applicable environmental laws and regulations, plus our own stringent environmental procedures."

In summary, a company needs to firmly decide what its environmental position should be. Does it aspire to be recognized as a regional leader in environmental performance? Or does the company policy call for aspiring to national industry environmental leadership as well as leadership in employee and public safety? Or third, does the company simply wish to maintain cost-effective compliance? Private codes can be an excellent vehicle for developing and communicating the company's goals relative to environmental position.

Performance Track Corporate Leaders

One other possible goal is the potential to be identified by the Environmental Protection Agency (EPA) as a Performance Track Corporate Leader. In 2004, the EPA created the Performance Track Corporate Leader designation as a device to recognize companies that exhibit environmental excellence in their policies and behavior at a corporate level and demonstrate substantial commitment to Performance Track. The Corporate Leader designation provides the EPA with additional opportunities to work more effectively with corporate leaders in improving environmental performance beyond regulatory requirements and is an opportunity for the corporate leader to provide a strong, positive influence and interaction with the EPA. The Performance Track Corporate Leader Program recognizes and promotes corporate activities not often or fully integrated at the facility level, such as improving the environmental performance of a company's suppliers or customers.

The criteria for designation as a Performance Track Corporate Leader are:

- Robust corporate management of environmental issues;
- Demonstrated environmental performance improvements and commitments to further improvements;
- Efforts to help improve the environmental performance of its value chain (including suppliers and customers);
- Corporate public outreach and environmental reporting;
- Strong overall environmental compliance record by the corporation, including its facilities that are not currently members of Performance Track;
- Plans to increase the corporation's level of membership in Performance Track and similar state performance-based environmental programs to at least 50% of its U.S. operations or at least 50 of its U.S. facilities within five years of designation as a Performance Track Corporate Leader; and
- At least five facilities of the corporation are each a member of Performance Track and similar state performance-based environmental programs represent at least 25% of its U.S. operations or at least 25 of its U.S. facilities.

Strategic Environmental Planning

Strategic environmental planning translates policy into a concrete framework for implementation. It includes assessment of the resources needed to achieve expected performance and documentation of goals, milestones, and performance expectations. Relative to the latter, it establishes the ways to measure achievement or lack thereof. Environmental planning also reflects appropriate segment-specific focus for the enterprise depending on the unique characteristics of the company's industry.

A company's Environmental Department is usually in large part responsible for the successful implementation of strategic environmental planning within the company. However, this can only be accomplished with effective direction and participation from senior management. It is senior management alone who can effectively tie strategic environmental planning to the overall business plan for the company. The more dynamic the competitive and regulatory climate, the more reason for senior management to consider a top-down approach to strategic environmental planning to ensure effective coordination across functional areas at the highest levels of the company. This may be accomplished through frequent, direct participation with the Environmental Department or the elevation of the senior environmental position to the executive management level.

Cross-functional coordination and cooperation are at the heart of strategic environmental planning processes and are essential to assuming that the strategic environmental process supports the core of the company's business plan. As part of the strategic environmental planning process, performance metrics should be developed and implemented to regularly assess both quantitative and qualitative aspects of the company's environmental management program. The environmental strategic plan should be available on the company's website and include metrics that have been established to measure the long-term performance of the environmental program.

In summary, as noted in Exhibit 18, the strategic environmental planning model should be an analytical model to evaluate whether the organizational systems effectively use the human and technology resources available, as well as adequately address the contingent technical, regulatory, and management issues.

Regulatory issues have profound effects on the organizational realities of environmental management programs, directly influencing technical and schedule and indirectly—but decisively—influencing cost. However, regulatory-driven organizational approaches are not necessarily optimal from the standpoint of long-term goals and performance objectives. Furthermore, from the company's standpoint cost is the key resource limitation and has a direct influence on the technical options, schedule, and ultimately the regulatory strategy necessary. Thus, there is a need for a strong, flexible environmental management organization and planning model that can take advantage of technical and schedule productivity opportunities and that can ease the divergent pressures of cost and regulatory forces, while in time doing the best to ease and minimize the divergence of the latter elements.

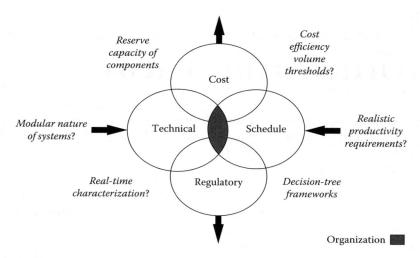

Exhibit 18. Strategic environmental planning model.

Reference

ISO. "Environmental Management Systems—Specification with Guidance for Use." Reference number ISO 14001:1996(E). (West Conshohocken, PA: ASTM. 1996.)

Chapter 7
Communication

The key criteria for the communication category are:

- External communication/public relations
- Internal communication
- Issue management systems

The focus of this part of the assessment is to measure whether the company's communication program achieves the following objectives:

Trust?   Clarity?   Proactive?

A consistent communications plan is a key to getting all employees mobilized. To do this, it must:

- Appeal to all employees' hearts as well as their heads;
- Assuage public concerns, both real and perceived;
- Provide for two-way communication, both with the public and employees;
- Address organization and individual project concerns;
- Permeate the organizational and public environment;
- Use a variety of media; and
- Provide an active role for management.

External Communication/Public Relations

External communication and public relations addresses the involvement with external constituencies such as environmental and regulatory bodies, elected officials, customers, and the public and includes public outreach programs, lobbying efforts, and intervention, where appropriate, in regulatory proceedings. Companies are increasingly being proactive, stepping up their external communications to promote their environmental position and, where applicable, outstanding environmental performance to stakeholders. This can be accomplished through the development of environmental reports to the community that rightfully take their place beside the company's annual financial report.

Companies are also increasingly using their Internet website to promote environmental performance and positions. But these efforts often need a focus because environmental information is often scattered in various locations and not given sufficient emphasis. According to recent surveys, almost 50% of the *Fortune* global top 250 companies are now issuing environmental, social, or sustainability reports compared to only 33% a few years earlier. The company's environmental reports must meet several crucial goals:

- Most importantly, it must be entirely a fact-oriented document. The company's reports should consist of concise descriptions of environment-enhancing programs. There should be no fluff and no unsubstantiated claims.
- The report must also be clearly written and well presented. It should be written in such a way that the lay reader can easily grasp the information, without fear of tripping over technical jargon or an overuse of acronyms. Well-chosen graphics and photographs are highly appropriate methods for conveying complex information and company environmental philosophies.
- The body of the report should fully support and be consistent with the message presented in the CEO's introductory letter.
- The report should present a comprehensive list of successful environmental projects as examples.
- Full regulatory compliance is an accomplishment important to all stakeholders. Reports should focus on this issue (or compliance trends) along with a discussion, if applicable, of the company's robust environmental audit program that ensures that compliance levels remain high.
- As more and more industries begin reporting on "toxic chemical releases" to the U.S. Environmental Protection Agency, under requirements of the Emergency Planning and Community Right-to-Know Act, it is useful to discuss reductions in toxic releases or future plans to achieve reductions.
- Information on oil and chemical spill trends, industrial health and safety performance, and energy conservation programs would all be useful additions to future environmental reports.

The report is an opportunity to establish a central theme and convey a "big picture" environmental stance that can inform the public of the strategic environmental tradeoffs that the company faces. In coordination with the company's public relations assets (internal and external), the environmental management should facilitate external communication initiatives such as enhancing the environmental communications online and doing systematic follow-up to the various outreach efforts such as environmental report and company literature.

Also, some companies have formed citizens' advisory groups to provide input to their environmental programs. Annual surveys are another way to assess the impact of the environmental program's public outreach efforts. Some other methods to enhance external communication are:

- Conduct an annual one-day "show-and-tell" program for regulatory agency personnel.
- Organize a Customer Advisory Panel to provide input on environmental components of company activities. The panel can be an asset for resolving community concerns prior to public meetings held by regulatory agencies in connection with environmental permitting actions.
- Set up a system where callers to the company's Customer Service Center receive EHS messages while they are on hold.
- Publish guidebooks on wildlife preservation for those areas relevant to the company's business.

The goal is to establish the company as an interacting, aware partner of the community on environmental issues and not a remote corporate entity.

From the perspective of international organizations, development of *corporate sustainable development reports* is very much in vogue. These reports have a broader reach than simply environmental performance, marrying economic and social development issues as well as environmental data and corporate commitments. The latter are increasingly being seen as both a necessary exercise to ensure that a "social license to operate" is maintained, as well as an opportunity to enhance brand reputation value.

In summary, as shown in Exhibit 19, effective external communication can draw upon linkages via regulatory relations, stakeholder towns, corporate environmental reports, and websites/telecommunications.

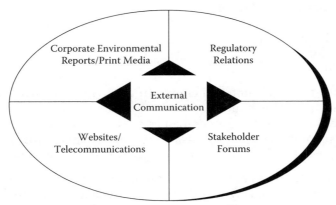

Strive for consistency in message and clear purpose behind the message

Exhibit 19. External communication links.

Internal Communication

Internal communication includes vertical communication to employees about company environmental goals and policies (and provides for employee feedback) and horizontal communication between the environmental functions and other areas of the company. Typically, both formal and informal communication networks fulfill the needs of the environmental function. Often, informal networks play a more critical role than recognized by management, but this informal network is very "people" dependent. The smaller the organization, the more susceptible it is to be dependent upon informal networks and to suffer loss in environmental performance during turnover. In these situations, the environmental services department must have sufficient formal networks and controls in place to successfully serve those organizations and facilities having environmental requirements.

Also recognize that even for large organizations, there is a risk of being too reliant on informal networks. The loss of key operational as well as environmental personnel can dramatically impact an informal network. In particular, the loss of environmentally cognizant line personnel can create a void in communication in certain situations and a consequent loss in the environmental function, at least until the replacement personnel are once again integrated into the network. Thus, mechanisms must be in place to either replace or develop capability.

An effective internal environmental communications plan may use multiple communications methods:

- Manager/employee meetings
- Live interactive satellite broadcasts
- Road shows (to the field)
- Video
- Newsletter
- Ideas form
- Hotline

It requires substantial time and energy from the Environmental Services Department to establish an effective environmental communication network. Increasingly, large-scale companies are relying on developing online training programs as well as accessing available training programs on the Internet. This enhances internal communication and integration.

Environmental messages can be communicated in numerous ways, including through monthly newsletters and even as "envelope stuffers" with paychecks. In addition, environmental program review presentations made at every quarterly board of directors meeting sends a strong internal communications message. Many companies find it helpful to develop a communications plan matrix that coordinates internal and external communication efforts. Exhibit 20 is an example.

Events/Channels: Project Week	Training Progress	Guidance Documents	SOPs	Internal Audits	News-letter	Q & A Updates	Fax Responses	Hot-line	Mgr./Employee Dialogues	Joint Team Videos	Pilot Videos	Evaluation Feedback Managers	Public Outreach	Community Functions	Live TV with Call-In
1									x						
2							x	x	x						
3	x				x		x	x	x						
4						x	x	x	x						
5		x			x		x	x	x						
6			x		x		x	x	x						
7						x	x	x	x						
8					x		x	x	x			x			
9				x			x	x	x	x					
10					x	x	x	x	x						
11							x	x	x						
12					x		x	x	x			x			
13						x	x	x	x	x					
14					x		x	x	x				x		
15							x	x	x		x				
16					x	x	x	x	x			x			
17							x	x	x	x					
18					x		x	x	x						
19						x	x	x	x		x				
20					x		x	x	x			x		x	
21							x	x	x	x					
22					x	x	x	x	x						
23							x	x	x		x				
24					x		x	x	x			x			
25						x	x	x	x						
26					x		x	x	x						
27							x	x	x						
28					x	x	x	x	x			x			
29							x	x	x						
30					x		x	x	x						
31						x	x	x	x						
32					x		x	x	x			x			
33							x	x	x						
34					x	x	x	x	x						
35							x	x	x						
36					x		x	x	x			x			
37						x	x	x	x						
38					x		x	x	x						
39							x	x	x						
40					x	x	x	x							
41							x	x							
42					x		x	x							
43						x	x					x			

Exhibit 20. Straw model communications plan matrix.

47

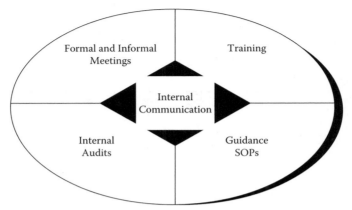

Both formal and informal networks can contribute

Exhibit 21. Internal communication links.

In summary, internal communication requires a strong interplay of training, guidance, and SOPs, formal and informal meetings, and internal audits (Exhibit 21).

Issue Management

Issue management addresses ways in which major environmental issues are identified, addressed, and resolved. This extends to "early warning systems" for new issues and ways in which accountability for issues management is assigned within the organization. Specifically, this element assesses whether the company manages major environmental issues proactively. This can be done primarily through the management's environmental management resources, which actively networks to establish relationships with environmental regulators at the municipal, county, and state level. However, on occasion senior management must be called upon to participate in major issue resolution.

The company's environmental management, with assistance from company lobbyists, must have in place a system to proactively track pending legislation that may impact company operations. Environmental management must also be prepared to successfully work at resolving the company's permitting issues as they arise. This often involves complex interdisciplinary skills (engineering, science, legal, regulatory) and effective management of contracting, consulting, and lobbying resources. For example, when negotiations stall environmental management may be forced to lobby its position to higher levels to successfully resolve permitting issues and save the company both time and money while maintaining environmental credibility. However, the well-established

relationship between the environmental management staff and the regulators can often prevent or minimize disagreements and controversies,
allowing issues to be resolved at lower regulatory levels with no negative
impact (i.e., fines).

Last, a company needs to decide when, how, and where it should play
a role on issue management. For example, a company may decide to play a
strong role within the state on issue management but decide for budgetary
reasons to restrict its national role to working through industry associations. One helpful technique to improve issue management is to form
internal or external committees to track emerging environmental issues.

Chapter 8
Functional Implementation

The key criteria for the functional implementation category are:

- Organization and staffing
- Internal integration
- Operational systems
- Risk assessment systems
- Waste minimization programs

The focus of this part of the assessment is to answer the question: "Does it translate into a culture?"

Organization and Staffing

Corporate management should at reasonable intervals examine environmental management organizational and staffing issues via a structured approach and take action to correct and update them as needed. Exhibit 22 provides a framework for analysis and implementation of organizational and staffing changes.

Organization and staffing should address the development of appropriate structural positioning and channels of communication for the environmental functions and the qualitative and quantitative adequacy of environmental staff resources at all levels and in all areas of the enterprise. It addresses whether the environmental staff and contractual resources are sufficient to meet milestones and goals. A deficiency here may or may not be reflected in a loss in end product performance; however, there may be concern that perceived strain on environmental resources may bode potential slippage in the future. In addition, there may be an issue that the environmental mission and requirements have increased substantially over the last few years and that holding at the established level is or may not be sufficient. This element also includes evaluation of environmentally aware line staff. For starters, there must be a current and accurate organization chart that clearly identifies the non-environmental line staff and their concomitant responsibilities, environmentally or otherwise.

This element also touches on human resources decisions regarding the reporting function of environmental personnel and their line of

				Implementation	
			Implementation	&	Institution-
Analysis	Macro-Design	Micro-Design	Planning	Development	alize
Create Under-standing of "As-Is" and Best Practice	*Develop and Select High Level Options*	*Design Detailed Elements*	*Develop Detailed Implementation Plan & Schedule*	*Implement & Develop the New Organization*	*Operation-alize New Organization*
Environment	Strategy	Structure	Leadership team development	Assume new roles	Stretch goals
History	Sucess factors	Roles and responsibilities	Implementation team development	Evaluation/ feedback	Reward systems
Internal	Model development	Governance			
Ongoing initiatives		Competencies/ skills		Extensive coaching	Extensive coaching
Drivers	Benefits and concerns	Selection process	Area wide "kickoffs"	Continuous improvement	Individual and organiza-tional learning
Design process	Option selection	Compensation/ people issues	Implement selection process	Build on the basic blocks	
Vision/strategy		Validation/ communication			Knowledge networks

Exhibit 22. Instituting ongoing feedback and learning.

administrative control over environmental policy and program control. A critical issue is the reporting function for plant environmental personnel vis-a-vis corporate environmental staff. The path leading to defining this organizational arrangement may involve several human resources decisions and may cost a firm financially as well as in terms of intellectual capital and morale. Human resource decisions should be made with consideration of the critical nature of environmental reporting at the operation level. Dedicated, well-trained personnel are needed to maintain the environmental reporting function with regulatory agencies, thereby avoiding any violations. Career paths and organization arrangements need to defend and enhance retention of key staff. Also, due consideration of vital environmental functions must be given whenever reorganization efforts occur.

In some cases, simply raising the level of the environmental program head to vice president can be a critical factor in integrating environmental activities throughout the company.

Internal Integration

Internal integration addresses the extent to which environmental goals, programs, and priorities are embraced and pursued by all elements of the enterprise and are reflected in planning and decision-making throughout (not just within the environmental functions). This addresses linkages between environmental and closely related health and safety programs.

As discussed earlier, a company is often dependent upon both formal and informal networks of communication to maintain the environmental function within and between line organizations. An environmental culture must be articulated throughout the units that overcomes the loss of personnel through turnover and reorganization. To consistently promote the integration of a company's environmental vision and policy throughout, the company's business units must be made to develop and conduct environmental training with the business units as a key to ensuring integration. Training positions and policies may be jointly shared with safety and environmental resource departments but it must clearly reflect environmental management needs.

As a part of internal integration, representatives from business organizations throughout the company should participate on environmental committees covering risk assessment and management, renewal of project licensing, spill investigations, and budgeting. Similarly, environmental personnel should participate in corporate committees or task forces on restructuring, marketing, strategic planning, and capital project planning.

A process should be developed to evaluate environmental management support practices and interaction with other operation and corporate groups. Exhibit 23 presents some examples of items to evaluate.

A company's record of environmental compliance is an indication that the management system—as well as the design, condition, and adequacy of existing operating equipment—continue to effectively meet environmental requirements and company expectations. In the case of continuous emissions and environmental compliance, it is wise for a company to establish a "safety factor" approach to avoid exceeding permit condition criteria.

A critical success factor to environmental compliance is the establishment by environmental management of *standard operating procedures* (SOP) for various environmental activities, including media sampling, chemical and physical analyses, property assessments, and so on. SOPs provide a consistent structure to activities, thereby enhancing quality, and also provide a means to orient and train new employees.

Operational Systems

Effective environmental management of operational systems addresses the scope and effectiveness of day-to-day environmental operations, provisions for dealing with unusual events and emergencies, and the operation and maintenance (including preventive maintenance) of major environmental facilities and equipment. In the past few decades, process systems

Health, safety and environment	Corporate management

■ Explicit knowledge of total costs (which are typically 2–6 time direct cost)
 	
■ Site accountability for
 	
  • SOPs/compliance
 	
  • Emergancy response
 	
  • Facility planning
 	
  • Community relations
 	
  • Health and safety
	■ Focus on analysis and information as well as data development
	
	■ Emphasis on environmental management partnership
	
	■ On-line, real time environmental performance reporting
	
	■ Integrated databases
	
	■ Management proactive in achieving desired environmental results
	
	■ IT environmental driven by line of business and plant operating strategies as well as regulation requests

Operations	Accounting
■ Environmental performance integral part of line managers' appraisals ■ Upward review/employee climate studies critical to environmental management performance assessment ■ Environmental programs and services tied to operating objectives ■ Shared services for economies of scale and skill ■ Line/environmental personnel rotation	■ Environmental technology/cost/rate of change decision methodology ■ Competition with outside service providers ■ Rigorous environmental buy versus provide analysis process ■ Continual benchmarking of environmental costs ■ Explicit project planning/execution process

Exhibit 23. Examples of functional implementation with other corporate groups.

have been increasingly designed to incorporate environmental operation systems. The latter are typically geared toward enhancing the safety of the operation, preventing and mitigating accidental spills and releases, and controlling and measuring for verification of regulatory accepted releases and emissions. The design and measurement standards of these environmental operational systems are geared toward prevalent air, water, RCRA, CERCLA, EPCRA, TRI, and NRC regulatory requirements.

Typically, the EPA's Resource Conservation and Recovery Act (RCRA) Facility Assessments (RFA) and EPCRA/Air/Water Permitting documents

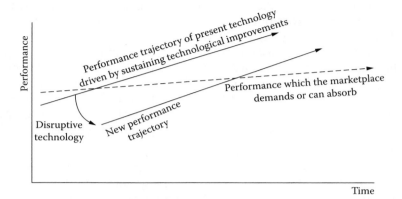

Exhibit 24. Disruptive environmental technologies: A need for leadership.

provide a strong profile of an organization's environmental operational system and its performance. In particular, Notice of Violations (NOV) can be scrutinized and used as one measure of the operational efficiency of the environmental systems. Additionally, industry background is available via the EPA's Industry Notebooks to assess industry operational trends from an environmental standpoint.

Environmental technology at the operations/plant level can often be seen as a disruptive technology. Management can sow the seeds of failure when confronted with the particular type of change. However, understanding and harnessing disruptive technological forces can often open very different market-based opportunities for alternative energy technologies and plant/product efficiencies, as illustrated in Exhibit 24.

The following are some examples of current environmental operational systems currently in place for several key industries and the potential problems that may occur.

Pollution Prevention Opportunities

Pollution prevention recognizes that the best way to reduce pollution is to prevent it in the first place. This often can be done by improving efficiency. This can lead to increased profits while at the same time minimizing environmental impacts. Among the ways in which this can be done are by reducing material inputs, re-engineering processes to reuse by-products, improving management practices, and employing the substitution of toxic chemicals with non- or less-toxic alternatives.

The EPA Office of Compliance, Office of Enforcement, and Compliance Assurance has compiled industry-sector notebooks that profile over a dozen industries' waste management results, methods, and operations.

Within each industry profile is a section that provides both general and company-specific descriptions of some pollution prevention advances that have been implemented for that industry and provides a starting point for facilities interested in beginning their own pollution prevention projects. This information may include a discussion of associated costs, time frames, and expected rates of return. However, note that facility-specific conditions must be carefully considered when pollution prevention options are evaluated, and the full impacts of the change must examine how each option affects the given facility's air, land, and water pollutant releases.

Industries currently spend a significant amount annually on environmental quality and protection. This alone provides the industries with a strong incentive to find ways to reduce the generation of waste and to lessen the burden of environmental compliance investments. Using the petroleum refining industry as an example, pollution prevention is primarily realized through improved operating procedures, increased recycling, and process modifications.

Both the EPA and industry have aggressively studied ways to implement more effective pollution prevention. For example, a cooperative effort by Amoco Corporation and the EPA studied pollution prevention at an operating oil refinery and identified a number of cost-effective pollution prevention techniques for the refinery. Also, the American Petroleum Institute has assembled a compendium of waste minimization practices for the petroleum industry based on a survey of its members. These are briefly described in Exhibit 25 through Exhibit 28.

In the case of petroleum refineries, it is often found that reducing pollution outputs through pollution prevention techniques also results in lowering operating costs. However, the primary barrier to most pollution prevention projects is at the start. Typically, pollution prevention options do not pay for themselves. Because corporate investments typically must earn an adequate return on invested capital for the shareholders, some pollution prevention options at some facilities may not meet financial requirements set by the companies. In addition, processing equipment is often very capital-intensive and has a very long lifetime. This reduces the incentive to make expensive process modifications to installed equipment that is still useful. But it should be noted that in the long run, pollution prevention techniques are nevertheless often more cost-effective than pollution reduction through end-of-pipe treatment. For example, the Amoco/EPA joint refiner study presents a case that the same pollution reduction currently realized through end-of-pipe regulatory requirements at the Amoco facility could be achieved at 15% of the current costs using pollution prevention techniques.

- **Place secondary seals on storage tanks**—One of the largest sources of fugitive emissions is storage tanks containing volatile products. These losses can be significantly reduced by installing secondary seals on storage tanks.

- **Establish leak detection and repair program**—Fugitive emissions are one of the largest sources of emissions. A leak detection and repair (LDAR) program consists of using portable detecting instrumentation to detect leaks during regularly sheduled inspections of valves, flanges, and pump seals.

- **Regenerate or eliminate filtration clay**—Clay from filters must periodically be replaced. Spent clay often contains significant amounts of hazardous waste. Back-washing spent clay with water or steam can reduce the content to levels so that it can be reused or handled as a nonhazardous waste.

- **Reduce the generation of tank bottoms**—Tank bottoms from storage tanks can constitute a large percentage of solid waste and pose a particularly difficult disposal problem due to the presence of heavy metals. Minimization of tank bottoms is carried out most cost effectively through careful separation system, including filters and centrifuges that can also be used to recycle.

- **Minimize cooling tower blowdown**—The dissolved solids concentration in the recirculating cooling water is controlled by purging or blowing down a portion of the cooling water stream to the wastewater treatment system. Solids in the blowdown eventually create additional sludge in the wastewater treatment plant. However, the amount of cooling tower blowdown can be lowered by minimizing the dissolved solids content of the cooling water. A significant portion of the total dissolved solids in the cooling water can originate in the cooling water makeup stream in the form of naturally occuring calcium carbonates. Such solids can be controlled either by selecting a source of cooling tower makeup water with less dissolved solids or by removing the dissolved solids from the makeup water stream. Common treatment methods include cold lime softening, reverse osmosis, or electrodialysis.

- **Control of surfactants in wastewater**—Surfactants entering the wastewater streams will increase the amount of emulsions and sludges generated. Surfactants can enter the system from a number of sources including washing unit pads with detergents; process system that produces spent caustics; cleaning tank truck interiors; and using soaps and cleaners for miscellaneous tasks. The use of surfactants should be minimized by using dry cleaning, high-pressure water or steam to clean surfaces.

- **Thermal treatment of applicable sludges**—The toxicity and volume of some deoiled and dewatered sludges can be further reduced through thermal treatment. Thermal sludge treatment units use heat to vaporize the water and volatile components in the feed and leave behind a dry solid residue.

- **Eliminate use of open ponds**—Open ponds used to cool, settle out solids, and store process water can be a significant source of VOC emissions. In many cases, open ponds can be replaced with closed storage tanks.

Exhibit 25. Examples of process or equipment modifications enhancements.

- **Remove unnecessary storage tanks from service**—Since storage tanks are one of the largest sources of emissions and release, a reduction in the number of these tanks can have a significant impact. The need for certain tanks can often be eliminated through improved production planning and more continuous operations. By minimizing the number of storage tanks, tank bottom solids and decanted wastewater may also be reduced.

- **Replace old boilers**—Older refinery boilers can be a significant source of SOx, NOx, and particulate emissions. It is often possible to replace a large number of old boilers with a single new cogeneration plant with emissions controls.

- **Reduce the use of 55-gallon drums**—Replacing 55-gallon drums with bulk storage can minimize the chances of leaks and spills.

- **Install rupture discs and plugs**—Rupture discs on pressure relieve valves and plugs in open-ended valves can reduce fugitive emissions.

- **Install high pressure power washer**—Chlorinated solvent vapor degreasers can be replaced with high pressure power washers which do not generate spent solvent hazardous wastes.

- **Refurbish or eliminate underground piping**—Underground piping can be a source of undetected releases to the soil and groundwater. Inspecting, repairing, or replacing underground piping with surface piping can reduce or eliminate these potential sources.

Exhibit 26. Examples of potential maintenance enhancements.

- **Segregate process waste streams**—A significant portion of hazardous waste arises from sludges found in combined process/storm sewers. Segregation of the relatively clean rainwater runoff from the process streams can reduce the quantity of sludges generated. Furthermore, there is a much higher potential for recycling from smaller, more concentrated process streams.

- **Control solids entering sewers**—Solids released to the wastewater sewer system can account for a large portion of a facility's sludges. Solids entering the sewer system (primarily soil particles) become coated. The Amoco/EPA study estimated that a typical sludge has a solids content of 5 to 30 percent by weight; preventing one pound of solids from entering the sewer system can eliminate 3 to 20 pounds of oily sludge. Methods used to control solids include: using a street sweeper on paved areas, paving unpaved areas, planting ground cover on unpaved areas, planting ground cover on unpaved areas, re-lining sewers, and cleaning solids from ditches and catch basins.

Exhibit 27. Examples of potential waste segregation and separation enhancements.

Unfortunately, a number of regulatory disincentives to voluntary reductions of emissions exist. As pointed out in the EPA's *Profile of the Petroleum Refining Industry* (September 1995):

> "Many environmental statutes define a baseline period and measure progress in pollution reductions from that baseline. Any reduction in emissions before it is required could lower a facility's baseline emissions. Consequently, future regulations requiring a specified reduction

- **Recycle and regenerate spent caustics**—Caustics used to absorb and remove contaminants from intermediate and final product streams can often be recycled. Spent caustics may be saleable to chemical recovery companies if concentrations are high enough. Process changes may be needed to make recovery of the contaminants economical.

- **Use nonhazardous degreasers**—Spent conventional degreaser solvents can be reduced or eliminated through substitution with less toxic and/or biodegradable products.

- **Eliminate chromates as an anticorrosive**—Chromate-containing wastes can be reduced or eliminated in cooling tower and heat exchanger sludges by replacing chromates with less toxic alternatives such as phosphates.

Exhibit 28. Examples of recycling and potential material substitution enhancements.

> from the baseline could be more costly to achieve because the most cost-effective reductions would already have been made. With no credit given for voluntary reductions, those facilities that do the minimum may be in fact rewarded when emissions reductions are required."

However, Clean Air Act amendments have increasingly encouraged voluntary reductions above the regulatory requirements by allowing facilities to obtain emission credits for voluntary reductions in emissions. These credits serve as offsets against any potential future facility modifications resulting in an increase in emissions. In addition, an aggressive trading market has evolved that makes it increasingly profitable to be "green."

Under the Clean Water Act, the discharge of water-borne pollutants is limited by National Pollutant Discharge Elimination System (NPDES) permits. Facilities that easily meet their permit requirements will often have their permit limits changed to lower values. However, because occasional system upsets do occur, resulting in significant excursions above the normal performance, facilities often feel they must maintain a large operating margin below the permit limits to ensure continuous compliance. Those facilities that can significantly reduce water-borne emissions through pollution prevention techniques have to weigh the risk of having their permit limits lowered. This can be a substantial disincentive.

Similarly, there is little positive incentive to reduce the toxicity of listed hazardous wastes because, once listed, the waste is subject to Subtitle C regulations without regard to how much the toxicity levels are reduced. Wastes failing a toxicity characteristic (TC) test are considered hazardous under RCRA. There is less incentive for a facility to attempt to reduce the toxicity of such waste below the TC levels because, even though such toxicity reductions may render the waste non-hazardous, it may still have to comply with new land disposal treatment standards under Subtitle C of RCRA before being disposed.

Environmental Risk Management

As discussed earlier, environmental risk management addresses the thoroughness, completeness, objectivity, and candor with which all types of environmental risks are identified and assessed throughout the company and programs developed to mitigate such risks.

Effective environmental risk management requires defining and capturing a future vision of environmental policy in current environmental planning activities. Limiting the vision to minimal current standards is a recipe for expensive retrofit and embarrassing public reassurance campaigns down the road. Successful environmental risk management also calls for understanding that environmental risk is financial risk as well as business, physical, legal, and political risk.

Many companies' risk assessment programs are informal and intuitive, except in unique circumstances. A formalized risk evaluation is comparable to what was discussed earlier in the Risk Oversight Committee discussions. The more formalized risk evaluation would be aimed toward assessing potential contaminant migration scenarios and the potential impact of various types of transportation accidents. The more formalized risk evaluation may be warranted dependent upon the scale and potential impact. The higher the level of concern, the more extensive the risk evaluation approach that should be taken. There is also recognition that the risks may be enhanced due to the increased operational activities and changes in the regulatory environment. A plan to proactively identify and abate increased operation hazards and regulatory risks may be warranted.

This may also include an internal survey of potential stranded assets. This involves a formal inventory, evaluation, and resolution of equipment and operations that may become stranded assets due to changes in operations and operating circumstances.

Building on the earlier discussions on the environmental review process, there are two steps to effective environmental risk management—assessing risk and shaping/exploiting the risk (Exhibit 29). The first step in risk management calls for identifying and assessing risk factors, prioritizing them, and profiling risk opportunities. There are two types of environmental management risks: "manageable" and "strategic." *Manageable risks* are characterized by a known environment where the problem and the capabilities and resources for redress are understood. In contrast, *strategic risks* deal with situations where the problems, capabilities, and resources are undefined and maybe even unknown. Strategic risks may call for major changes in market, process design, or organization. The next step is to shape the risk by quantifying and controlling the financial implications. It may require allocation of capital or a shift in the strategic direction of the firm.

Phase I—Assessing Risk (From All Sources)

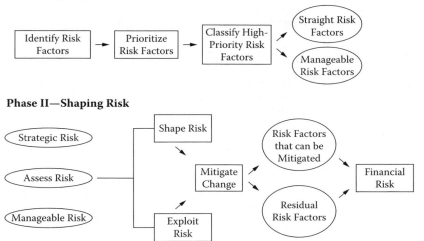

Phase II—Shaping Risk

Exhibit 29. Pathways to environmental risk management.

There must be an effort to exploit the risk by analyzing opportunities and developing plans to implement desired changes. To do so, a framework must be developed for ongoing monitoring and self-assessment to ensure ongoing risk management is successfully effected.

The key steps to this process are to:

1. Develop an exhaustive list of potential risk scenarios.
2. Define the risk threshold the company is willing to accept for these scenarios.
3. Quantify the risk threshold for these scenarios.
4. Rank the order of these scenarios.
5. Begin a classification effort and preliminary plan to manage the risks.

Inherent with the effective implementation of a risk management process is the question, "To where does it report?" Establishment of an Environmental Risk Management Committee at the board of directors' level ensures not only timely implementation of the risk management process but also timely response to its findings.

Last, to support this approach, key environmental personnel should receive root-cause analysis training.

Waste Minimization Programs

Waste minimization addresses programs to develop and implement activities dealing with waste reduction, pollution prevention, and recycling. This

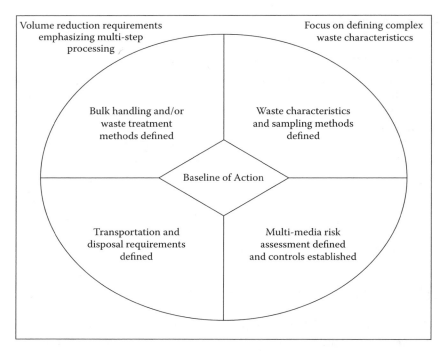

Exhibit 30. Functional implementation criteria model.

element assesses whether the company has developed a formal pollution prevention program for the entire organization and what level of priority and implementation has occurred. A high priority is warranted, given that these programs are a primary method for generating savings and possibly revenue for the company. However, to be successful it is critical to establish an organizational structure and business process to systematically implement the pollution prevention program. In addition, the program needs a committed champion among senior management to facilitate its development and implementation. In conjunction with the rollout of a pollution program, a recycling program should be developed as a revenue-generating operation, if economically feasible.

The functional implementation criteria model is the analytical tool to evaluate each given "baseline for action." As shown in Exhibit 30, this process evaluates the given management's readiness to mesh and move forward in its environmental management program from an operational system standpoint.

The step of the internal assessment process will identify critical gaps and inconsistencies in environmental management practices within the organization. Furthermore, this step provides a baseline for evaluating the potential impact of other technology development options that may be on the horizon.

The Emission Reduction Program

An increasingly critical component to assessing environmental management effectiveness is emissions reduction. Most firms are capable of generating metrics to reflect up-the-stack emissions but few have accepted the challenge to a comprehensive baseline reading. California and the World Resource Institute's Greenhouse Gas Protocol initiatives provide an accepted standard for establishing corporate baselines.

The identifying sources step of the review involves specifically identifying sources and potential impacts of sources. Emission sources may be categorized as follows:

- Stationary combustion emission sources
- Transportation-related mobile emission sources
- Indirect stationary emission sources
- Process stationary sources
- Fugitive stationary emission sources

As part of the source identification process, specific greenhouse gases of concern relative to each source type must be identified and the potential impact established. These include:

- Carbon dioxide (CO_2)
- Methane (CH_4)
- Nitrous oxide (N_2O)
- Hydrofluorocarbons (HFCs)
- Perfluorocarbons (PFCs)
- Sulfur hexafluoride (SF_6)

The confirming emission data step entails implementing procedures for checking the accuracy of greenhouse gas emission data. It draws upon available documentation.

For stationary combustion sources (i.e., power plants, refineries, manufacturing facilities), this involves confirming fuel type consumption patterns, emission calculation factors and assumptions, as well as continuous emission monitoring systems (CEM) accuracy and maintenance (calibration, QA/QC). Sampling techniques may be used to supplement and statistically confirm data development.

Transportation, indirect, process, and fugitive emissions are developed by using estimation techniques based on utilization and production records, fuel consumption records, and a review of relevant maintenance histories. In some cases, sampling techniques may be used to confirm estimation assumptions.

For all sources, greenhouse gas emissions are calculated in mass terms. Within the U.S., procedures to calculate combustion-related mobile sources can be based on methods presented in the *Inventory of California Greenhouse Gas Emissions and Sinks: 1990–1999* (CED, 2001).

Sources for emission factors may include:

- U.S. EPA. "Compilation of Air Pollutant Emission Factors AP-42." *http://www.epa.gov/ttn/chief/ap42.*
- U.S. EPA. "Emissions Inventory Improvement Program (CHIIP), Introduction to Estimating Greenhouse Gas Emissions: Volume VII" (EIIP, 1999). *http://www.epa.gpv/ttn/chief/eiip/techreport/volume08/index.html.*
- IPCC. "Revised 1996 IPCC Guidelines for National Greenhouse Gas Inventories, Greenhouse Gas Inventories Reference Manual" (IPCC, 1996).
- U.K., Department for Environment, Food, and Rural Affairs. "Guidelines for the Measurement and Reporting of Emissions in the UK Emissions Trading Scheme" (DEFRA, 2001b).

Direct emission factors include a factor to reflect fuel mix as well as a transportation and distributing loss factor.

The final step is to evaluate and report findings. The final report specifies whether the organization or facility of concern falls within the accuracy requirements called for by the registry (i.e., in California it is 5%). Greenhouse gas emissions may be presented in absolute terms or on a normalized basis (i.e., divided by some agreed-upon entity output).

Evaluation and report findings done in accordance with California reporting guidelines are per the requirements of SB 527. The latter was developed based on review of WRI/WBCSD GHG protocol (WRI, 2001), the Australia Greenhouse Challenge Program (AGO, 1999), the U.K. Emissions Trading Program (DEFRA, 2001a; 2001b), the U.S. DOE 1605b voluntary program (EIA, 1994; 2001b), the U.S. EPA Climate Leaders Program (EPA, 2002), and Canada's Climate Change Voluntary Challenge Program (VCR, 1999).

As a final note of encouragement toward this process, cutting emissions has less to do with do-gooder ambitions and more to do with energy efficiency and direct bottom-line results. Many firms have found that revamping their businesses in a resource-efficient way not only reduces greenhouse emissions but cuts operational costs, too.

Chapter 9
Measurement Systems

The selection of metrics begins with objectives. Objectives determine goals to be achieved for the environmental management function as well as the users and the mission context. The methods for developing environmental metrics are straightforward: define the strategies, processes, and organizations put in place to accomplish objectives and select metrics that indicate how well you are performing against your objectives. As shown in the Metrics Development Wheel, there is a wide variety of elements to potentially draw upon.

Besides regulatory performance (notice of violations, and so on), other key criteria for the measurement systems category are:

- Audit program
- Environmental management information system (MIS)
- Cost and financial management

However, as shown in Exhibit 31, there is an array of core competencies that can be identified for measurement.

The key questions to answer are: Do the environmental costs and emission/release data effectively measure the entire organizational environmental performance? and, Do the environmental metrics reflect the company's core competence attributes?

Audit Program

Among the tools that assist companies in implementing their environmental management systems (EMS) are an environmental impact assessment, environmental accounting, and auditing and life cycle assessment. Auditing in particular is a powerful tool for assuring company managers of accurate information and also contributing to the external credibility of a company's environmental commitment.

A sound audit program is a critical part of a successful EMS. An audit program should address the comprehensiveness and effectiveness of environmental auditing activities, including the setting of standards (e.g.,

Exhibit 31. Metrics Development Wheel.

compliance or "beyond compliance"), the definition of protocols, the involvement of line management, and the implementation of results.

Key questions include: Has an environmental auditing program been established? Does the audit program ensure that all its facilities are audited on an annual basis? Has a formal auditing baseline record been established for each facility? In addition, the company should develop environmental inspection and self-audit forms for use by the line organizations covering all relevant regulations and company procedures.

Besides a broad but limited depth facility audit approach, one can consider blending an approach that targets a smaller number of audits but with a more in-depth nature, looking more carefully at a given facility, conducting equipment checks as well as collecting and analyzing samples to verify facility records. The scope of auditing activities should also include environmental services department programs, environmental training, and emergency response. Another way for the company to enhance its audit program is for the company to participate in joint vendor audit programs and external peer organizations dealing with comprehensive environmental self-assessment programs.

Typically, a first reaction is to use the "audit approach" to tighten controls. However, a word of caution about controls—they are not the be-all and end-all! Solely emphasizing conventional "audit approach" control methods without marrying them to an understanding of the corporate and industry risk cultures invites unnecessary inefficiencies with subsequent cost, schedule, and production implications with little risk management gain.

The EPA is on the cusp of developing a working definition of EMS and third-party certification for enhanced environmental performance. Much of the focus has been oriented on federal facilities, given the Clinton Executive Order for EMS development by federal facilities by 2005–2006. Yet there is increasing thought to expanding the EMS review process to include industry in general. One would think that industry would be resistant to such a movement but there are some in industry who are seeing it as a mechanism for providing enforcement inspection relief to those corporations who demonstrate strong EMS programs.

An EMS allows an organization to systematically manage its environmental and health safety matters. The EMS process is a continual cycle of planning, implementing, reviewing, and improving the processes and actions undertaken to meet the organization's business and environmental goals. Most EMSs are built on the "plan, do, check, act" model that leads to continual improvement based upon:

- The *plan* element, including identifying environmental aspects and establishing goals through planning;
- The *do* element addresses implementation activities, including training and establishing operational controls;
- The *check* element includes monitoring and corrective action needed to ensure implementation; and
- The *act* element includes progress reviews and acting to make needed changes to the EMS.

The EMS can result in both business and environmental benefits. For example, an EMS may help a company:

- Improve environmental performance
- Enhance compliance
- Prevent pollution and conserve resources
- Reduce and mitigate risks
- Attract new customers and markets
- Increase efficiency
- Reduce costs
- Enhance employee morale
- Enhance their image with public, regulators, lenders, and investors
- Achieve or improve employee awareness of environmental issues and responsibilities

- Qualify for recognition or incentive programs such as the EPA Performance Track Program

There are costs that the latter may include:

- Investment of internal resources, including staff and employee time
- Costs for training of personnel
- Costs associated with hiring consulting assistance, if needed
- Costs for technical resources to analyze environmental impacts and improvement options, if needed

EMS and third-party certification are opportunities to drive forward environmental performance and to address the source of increasing risk in business. ISO 14001 is a checklist of needs for good environmental management. The traditional approach to EMS implementation is certification of limited reach—documenting system conformance and establishing whether practices conform to documented procedures. It focuses on demonstrating compliance (to the standard) and is centered on the documented system manual. However, this approach is not consistent with the way people (managers) get results. The thought process behind the EMS program approach is that environmental performance outcomes arise from management and operator behaviors. EMS, auditing, and certification is an opportunity to enhance performance by harnessing management commitment (to certification and successful audit outcomes) and the resources that come with it.

Given the focus on changing management and operator behaviors, senior management often will use the auditors to support their efforts to obtain and maintain throughout the organization while driving toward EHS goals. The goal of auditing management and operator behaviors and outcomes to drive improvement should not be to either soothe management and relieve them of responsibility nor beat up on management, environment personnel, and operators. The goal should be to establish current practices, to help the organization recognize the consequences (good and bad) of current practices, and to make management feel accountable for fixing them. The latter should be a strong concern of the board of directors. Auditing management and operator behaviors and outcomes to drive improvement should include the following: site tours, root cause analysis, business performance review, environmental management, meeting with the top person, and a closing meeting.

Site Tours

During site tours, the assessment should be centered on performance outcomes by focusing on:

- Potential accidents and incidents
- Non-compliance
- Ongoing impacts

Again, the philosophy is that good systems generate good outcomes; poor outcomes generate system weaknesses. However, testing by simply seeking examples of poor outcomes often proves to be a weak assessment that fails to find deficiencies that are developing.

Site tours should focus on key discharges, ultimate discharge points, chemical handling and storage, lifting gear, drainage, roof work, fire control, and contractor activities. Auditors should photograph every issue they see. This includes sources, pathways, receptors, and consequences for them. Photos provide focus and can be used for training and the closing meeting. It also helps eliminate argument and focuses action on the ground.

Root Cause Analysis

Everything seen during an assessment is material evidence of operator behaviors, the extent of management engagement in driving the environment, and the effectiveness of programs established by management. For the selected issue or fact, the auditor must seek to understand—through discussion with escorts, operators, supervisors, and managers—the operator behaviors that have given rise to the observed outcome or fact, why operators have failed to do what they should have done, or why they have done the wrong thing (i.e., establishing the cause of the occurrence).

As each issue comes to light, the auditor must focus their attention to understand why this has come to pass and how it relates to other occurrences. The auditor must never make assumptions but seek only explanations for what is discovered. The questioning style should be open, indirect, and never leading. It is preferred for the auditor to have knowledgeable escorts and to solicit their input throughout the audit. Each audit element should take three to four hours (or more), so carefully select the issue for analysis via agreement of the problem or concern to be reviewed and the crossed boundaries (shifts, departments, relevant to many) that must be canvassed. Auditors should write up findings in a manner that clearly establishes facts, consequences, immediate and root management weakness to be addressed, and the standard to be met.

Business performance reviews are geared toward understanding the context of operations and how it will influence the environment. The business performance review should establish where the company sees their greatest strengths in mainstream activities and what business status and developments might impact their environmental program. This analysis can be conducted at many levels: for corporation as a whole, for a division, for a site, or for a single department. Key areas of the business performance review include performance trends (revenues, profits, and margins), personnel, competitive strengths, customers, extent of local control over operations, output versus capacity, products and anticipated changes in them, development plans, and on other exciting developments of note.

The environmental management review itself is a top-down analysis. It seeks understanding of the techniques management uses to drive the operations and the extent to which management uses (or does not use) these techniques to implement environment management goals. The environmental management review should include facts from discussion with management and the site tour, consequences, immediate and root management weakness to be addressed, and a definition of goals and recommended standards to be met.

The report should be specific, focusing on plant, people, processes (real ones), and practice. The report must clearly discuss consequences for the business and its management. Can management be comfortable with the way they appear? To do this, the report must identify main causal factors to ensure that the management team understands management causes and establishes accountability for fixes. Typically, an environmental function in most organizations can identify and oversee (i.e., facilitate) but it cannot implement.

When it is all said and done, EMS certification criteria come down to the following questions:

- Is there evidence that major potential incidents are not properly controlled?
- Have effective mechanisms been established to prevent accidents and incidents from occurring?
- Is there evidence of any major regulatory non-compliance?
- Have effective mechanisms been established to deliver compliance, and are they currently delivering compliance?
- Are ongoing impacts on people and the environment giving rise to major concerns?
- Have effective mechanisms been established to control ongoing impacts, and are they yielding control?
- Have mechanisms been established to deliver ongoing reductions in impact on people and the environment?

Good audits will reveal reality, induce serious management engagement and behavioral change, and deliver enhanced outcomes. Good audits should drive sustained performance improvements (better outcomes) and result in:

- Fewer accidents and incidents
- Higher levels of compliance with regulations
- Reduced impact on people and the environment

Environmental Management Information System (MIS)

Environmental management information systems (MIS) addresses the availability of easily accessible online systems to track environmental

Key Performance Measures	
Measure	**Definition**
Regulatory performance	Meeting permitting and reporting deadlines
Materials index	Waste minimization effectiveness
Capacity utilization	Actual production to capacity
Operations performance	Equipment reliability, accuracy of emissions data, release data (discharges, emissions, spills)
Maintenance indices	Maintenance costs and performance levels (routine and turnaround)
Capital investment	Capital reinvestment rate
Total equivalent personnel	Head count per function/activity including contractors
Monthly expenditure	As a percentage of revenue

Exhibit 32. Examples of potential environmental management information system (MIS) elements.

performance, activities, and issues. Exhibit 32 provides some examples of items that might be included as part of an environmental MIS.

Establishing an MIS to track waste streams facilitates better management of waste generation. The results of a successful MIS program are typically significant, in some cases resulting in a 90% reduction in wastes generated. However, after achieving waste reduction at the outset, often use of this waste-tracking MIS may languish. An ongoing emphasis must be maintained and upgrade reviews conducted on a timely basis to ensure that the environmental MIS continues to be actively used once installed. There should be a continuing vigilance to identify and evaluate new waste management software in an effort to continually enhance consistent waste tracking throughout the company's operations.

Under this element, a company's historical performance in meeting critical permitting and reporting deadlines to environmental regulatory agencies should also be assessed. These deadlines should be tracked consistently across the organization. Notices are generally sent out by the regulatory agencies, but implementation of an MIS to track regulatory permitting and reporting deadlines facilitates the management of environmental staff time. This is particularly critical during crunch times when multiple deadlines occur within a relatively short period. The plant staff could also make use of the MIS to establish a maintenance notification system, which could be used to send out reminder notices to conduct inspections, send reports, and so on. Ultimately, having a tracking system in place will free up time and facilitate changes in personnel because information will not have to be personally transferred to ensure that the new employee knows when deadlines are coming.

Last, effectively managing sample and analytical data collections actively, in and of itself, can be a challenge. Regulatory agreements specify monitoring activities that often enter into a "black hole" when it comes to effectively summarizing the data in a meaningful way that facilitates corporate remedial planning or legal defense positioning. Different sites and different laboratories may generate different data set protocols that are not easily merged and get locked away into hard files of the "informal" network.

Environmental Cost Management

Whereas we recognize that tougher competition leads to greater bottom-line focus and feeds a tendency to view EHS activities only as cost, leading companies know how to manage EHS such that EHS is more than just cost but can be viewed as a business asset.

It is typically found that traditional itemized EHS costs can be 6% of sales, but total environmental costs may be two to four times higher when hidden costs are factored into the analysis. The latter may include costs relative to non-EHS staff time, environmental implications on capital asset productivity, and additional new product development or liability costs that may be incurred due to environmental concerns. Without proper control, these costs may cause unwarranted loss of share value and severely impact revenues and earnings. Many investors feel that these types of non-financial EHS costs will become more important to the bottom line over the next five to ten years and will pay a premium for companies whose EHS strategies lead to effective control and competitive strategic advantage.

Despite current U.S. policy, there is an increasingly aggressive effort within the U.S. to more accurately reflect environmental liabilities on company 10-K reports pursuant to Rule 192 of the Securities and Exchange Commission (SEC). This of itself is consistent with and is a basic building block for sustainable development. The issue was flagged by a 1993 Government Accounting Office (GAO) study and a 1998 EPA Office of Enforcement and Compliance Assurance study that found significant deficiencies in 10-K reporting of environmental liabilities. The debate centers on industry's inability to estimate potential environmental liabilities due to "uncertainties" and, in some cases, a failure to identify the "aggregate" liability that may trigger materiality but is not recognized on an individual claim basis.

Thus, a greater emphasis is warranted toward tracking actual environmental costs. These costs are typically higher than the values provided

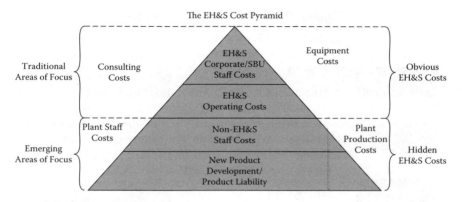

Exhibit 33. The EHS cost pyramid.

through typical accounting systems. Activity-based accounting would provide a more effective means of tracking environmental costs than the typical financial tracking system currently in place.

ASTM International has entered the debate with its issuance of the "Standard Guide for Disclosure of Environmental Liabilities" (EZ173-01) and the "Standard Guide for Estimating Monetary Costs and Liability for Environmental Matters" (EZ137-01). There is an increasing emphasis on strong statistically or actuarially based approaches to establishing environmental liabilities, and a strong emphasis on an independent estimation and verification by well-qualified independent reviewers. Effective environmental cost management addresses the availability of accounting systems for environmental expenditures of all types and a system to track overall expenditure trends, including capital and operating and maintenance (O&M) charges.

As part of an organization's effort to identify and refine environmental-related accounting, in 1999 the Ricoh Group built an environmental accounting system as a managerial decision-making tool. Ricoh's Segment Environmental Accounting and Business Sector Environmental Accounting are used as internal accounting tools, as well as corporate environmental accounting, and are geared toward defining and promoting sustainable environmental management. Further upgrades in the system are used in mapping out environmental action plans, selecting measures, and confirming achievements.

In this way, environmental accounting is used to determine measures to promote sustainable environmental management. Environmental impacts are reduced by using measures that will lead to the creation of benefits and to promote sustainable environmental management. The Ricoh Group uses environmental accounting to help determine what measures should be

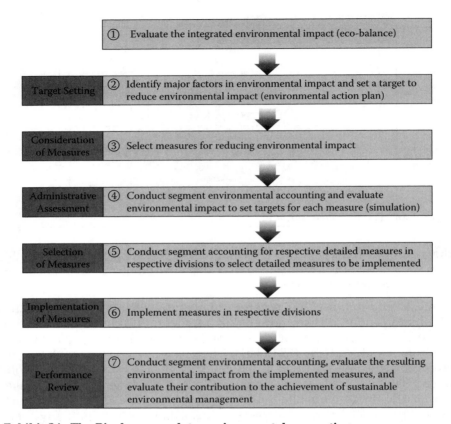

Exhibit 34. The Ricoh approach to environmental accounting.

taken for what processes and for what operations so that the maximum effect can be obtained. The first step is to identify those processes that have a high environmental impact in business operations, based on Ricoh's Eco Balance model. One then examines a number of improvement plans to reduce the identified environmental impact, in consideration of developments in society, laws and regulations, and competition. Then, using segment environmental accounting the analyst assesses the effectiveness of each possible approach and decides what methods should be adopted to gain the best results. Ricoh's approach is laid out in Exhibit 34.

Corporate environmental accounting is a tool to inform the public of relevant information compiled in accordance with the Environmental Accounting Guidelines of Japan's Ministry of the Environment. The Ricoh Group approach is to take the necessary portion from the Eco Balance data and calculate the cost and effect (in quantity and monetary value) of its environmental conservation activities based on its own formulas and indicators. The calculated results are disclosed to the public after being

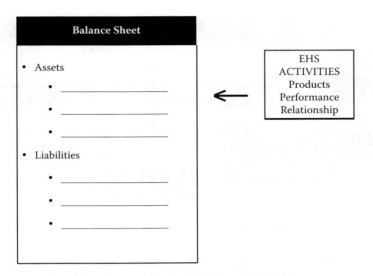

Exhibit 35. Strategic management of EHS investments.

verified by a third-party organization. The ultimate goal is to make the corporate Environmental Accounting Statement comparable to already standardized documents, such as financial statements.

Segment environmental accounting is an internal environmental accounting tool. It is used to select an investment activity or a project related to environmental conservation from among all processes of operations, and to evaluate environmental effects for a given period. The effect of investment on environmental conservation is calculated based on the accepted financial concept of "return on investment" (ROI). It is recommended that the calculation result be used internally for decision-making sustainable environmental management and not be part of an external reporting process. Companies and divisions increasingly use segment environmental accounting for their operations.

Business sector environmental accounting is a service provided by Ricoh. It is a broad indicator of how various environmental activities contribute to environmental management conditions in respective business sectors.

Environmental Asset Management

As shown in Exhibit 35, typical corporate accounting systems only capture the superficial environmental assets. To manage EHS assets as critical business assets, a value chain must be established based on identifying EHS assets, assigning values, and managing assets strategically. An example of a value chain is provided in Exhibit 36. As shown in Exhibit 36, value may stem from performance, product, or relationship benefits. For example,

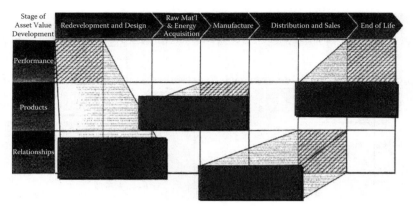

Exhibit 36. Example of an environmental asset value chain.

good community relationships enable the expedited approval of plant expansion. "Green" energy production value may differentiate the company's commodities in the market and generate product loyalty. And finally, a recycle infrastructure may enable performance-based cost savings.

It should be recognized that in many industries, EHS may be a strong limiting factor in aggressive utilization of operational capacity if not well-managed from an operational standpoint. However, recognize too that recent developments in national air emission trading schemes and land banking have provided industry with an opportunity to translate strong EHS policies into tradable assets. This is particularly relevant from a global perspective where such practices have been well established. Recognize that the asset generation implications of emission reduction and consequent emission trading credit development are not minor. In the 1990s, global trading of emission credits doubled from $4 trillion to $8 trillion and, despite the U.S. rejection of the Kyoto Protocol, will continue to increase. In fact, much of the current administration's "Clear Skies" policy (in lieu of Kyoto) hinges on a thriving emissions trading market.

So depending on the company, including environmental allowance emission credits may be extremely beneficial. As such, there should be a strategy to manage these credits because they represent an environmental asset. These environmental credit assets may include additional air-related assets (e.g., NO_x credits), water-related assets (i.e., credits under the NPDES program), and land-related assets (e.g., wetland and habitat banks). These current and future assets should be identified and characterized. Plans should be developed for managing these assets, including accumulation and asset sales strategies.

Last, while EHS may have a negative impact on operational asset utilization, more efficient management of the EHS issue in comparison to the com-

Exhibit 37. A capital value approach to establishing value of performance.

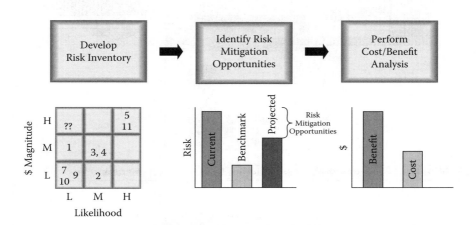

Exhibit 38. A risk-based decision approach to valuing performance assets.

petitive may also provide a significant asset edge. One can use a capital value, a risk-based design, or a stakeholder satisfaction approach to establishing value. Exhibit 37, Exhibit 38, and Exhibit 39 present graphical snapshots of these approaches. As shown in Exhibit 40, further analysis can be done to classify the competitive position impacts of EHS asset strategies.

The objective is to provide investment analysts with quantifiable environmental information that can be translated into hard financial data. The latter can be accomplished through discounted cash flow analysis, risk analysis, or real option analysis techniques. Real option analysis may involve establishing the firm's competitive advantage as defined by brand reputation value and its impact on the "social license to operate." This may be defined in terms of lengthening operating timeframes or identifying "green" pricing advantages.

Once expected results (benefits) and costs are established, a portfolio approach can be taken to break assets into excitement, performance, and threshold assets (Exhibit 41).

Environmental Financial Management

Evaluating alternative financing instruments for environmental projects is proving to be a critical factor in ensuring optimal asset utilization within a

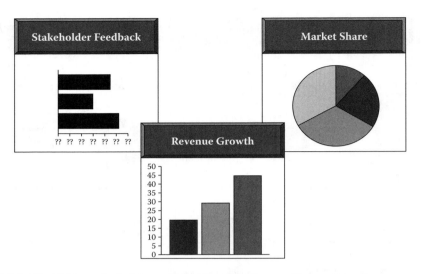

Exhibit 39. Using stakeholder satisfaction to value excitement assets.

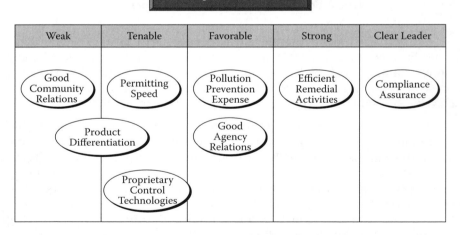

Exhibit 40. Establishing the competitive position of assets.

sustainable development framework. As environmentally related project financing has grown in recent years, so too has the array of financial investments tapped to meet the special borrowing needs of industry. The choice of financial instruments varies, dependent upon the type of environmental project involved and the potential for identifying tangible asset utilization efficiencies and asset life extension benefits.

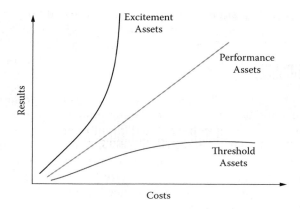

Exhibit 41. A portfolio approach to valuing EHS assets.

Given the current fiscal strain within industry segments, the most intriguing categories of project financing are off the balance sheet, free of recourse to the sponsor. However, off-balance-sheet financing constitutes a clear environmental liability on the part of off-balance-sheet third parties that are interested in the project's completion. Typically, these instruments are private debt issues, arranged through a commercial or investment bank and adapted to the peculiarities of environmentally related project financing.

The following are some examples of possible financial instruments and funding sources for environmentally related projects.

Conventional Commercial Loan

Conventional commercial loans are the largest source of environmental project financing and may take the form of either secured or unsecured loans. This type of loan typically covers the more well-defined type of environmental project activities, such as development of well-tested treatment, storage, and disposal concepts for hazardous waste material handling. These loans may be in the form of construction loans, term loans, bridge loans, mortgage loans, or working capital loans, dependent on the specifics of the proposed environmental project. However, commercial banks tend to limit their commitments to five to ten years with fluctuating rates based on the U.S. prime rate. Also, conventional commercial bank loans offer little opportunity for off-loading financial costs and risks in lieu of potential gains in asset efficiency, utilization, and life extension.

Supplier Financing

Supplier financing and associated captive financing companies are often a good source of funds for high-end, sophisticated environmental technology project financing. Many manufacturers of large, environmentally related capital equipment are increasingly developing captive financing arrangements to assist buyers in arranging the financing for their products via secured and unsecured term loans, as well as installment loans and leases.

Another innovative project financing approach is to negotiate production payment loans and advances. These instruments are widely used for both off-balance and on-balance-sheet financing of energy and manufacturing related environmental projects with the obligation being to repay from the proceeds of enhanced production. An example would be expanding a natural gas pipeline connection to eliminate flaring practices. Not only is there a benefit from proceeds of the returned gas but also from an emission credit trading standpoint as well.

Commercial Paper

Commercial paper is typically accessed for short-term financing but can be rolled forward to provide long-term financing needs as necessary. Environmentally related project financing with significant unknowns can tap into this market if backed by a letter of credit or line of credit from a commercial bank or insurance company. The letter can make the commercial paper alternative an attractive option compared to other methods of financing when speed of response is critical (i.e., major spills).

Bond Financing

Bond financing creates special problems for most environmental project financing. Access to the public debt market is bound by both federal and state securities laws that may hamper the flexibility often required by environmental financing. However, sponsors and guarantees with established credit and well-defined environmental project scenarios are able to access the public markets using instruments that take advantage of:

- Securities Act Rule 415, which permits easier, quicker, and more flexible access to the debt markets than bond instruments of the past;
- Securities Act Rule 144A, which eliminates the two-year holding period previously imposed on buyers of privately held securities; and
- Registration Form S-3, which relies almost entirely on incorporation by reference to exchange act reports and a basic prospective development.

For example, an environmental medium-term notes (MTN) issue can be floated over all or part of the life of the security with a coupon reset formula based upon an agreed benchmark. Typically, these are asset-backed securities. The motivation for segregating environmentally related assets as a collateral offering is that it can result in lower funding costs, depending on the credit quality of the issuer and the trade or usage flexibility of the environmentally driven project asset. However, recognize that it is unlikely an individual bank would be able to secure a major environmental project finance loan and, subject to size, lending consortiums may need to be created.

Private Placement Debt

Private placement debt differs from public offerings in that private placement:

- Does not require regulatory approval;
- Does not require public disclosure; and
- Is arranged with a limited number of sophisticated institutional investors.

However, recognize that private placement debt may be accessed through commercial or investment banking venues.

Among the benefits of private placement debt to borrowers is that they can retain absolute control of the placement and can establish long-term relationships that may prove useful for future environmental financing. Also, the fact that private placement debt requires no public disclosures of sensitive information has the advantage of allowing environmental redress to be conducted with the minimal public scrutiny required by environmental regulations. Whereas private placement debt interest rates may be higher than U.S. market rates for a similarly rated debt, this may be offset by establishing production schedule "takes" in lieu of more straightforward financing arrangements.

Environmental Capital Equipment Leasing

Environmental capital equipment leasing constitutes an excellent source for both on-balance-sheet and off-balance-sheet financing. These leases can be either tax-oriented or non-tax oriented. In the tax-oriented lease, investors may claim and retain the tax benefits associated with the environmental equipment ownership while passing the benefits to the lease in the form of reduced payments. For companies that are not currently generating sufficient earnings to cause income tax liability, a tax-oriented leasing strategy offers the opportunity to indirectly obtain tax benefits associated with environmental equipment ownership that would not be available at the time the equipment was purchased.

This type of environmental lease can be structured as a leverage lease in which the equity investor finances a portion of the funds (20%) and lenders provide the balance needed to acquire the environmental asset being leased. In this case, the lenders' security interest in the leased asset would be senior to the equity interest. However, the latter may enjoy tax benefits by claiming tax deductions and tax credits, if applicable, upon the entire cost of the asset.

Environmental Related Preferred Stock

Although not a debt instrument, environmental related preferred stock has the characteristics of both common stock and debt and enjoys the advantage of being attractive to the investment community with specific targets for "green" investments. Similar to the holder of common stock, the environmental related preferred stockholder would be entitled to dividends but at a specified percentage of par or face value. Unlike debt, recognize that environmentally related preferred stock payments would be treated as a distribution of earnings and whereas interest payments are tax deductible to the corporation, the dividends are not. However, the U.S. tax code exempts 70% of qualified dividends from federal income taxation if the recipient is a qualified corporation. Thus, if the major buyers of the environmentally related preferred stock are corporations seeking tax advantages, the latter may be transferred via acceptance of a lower dividend rate.

Master Limited Partnerships

Master limited partnerships are another financial device to fund environmental project activities. Under such partnerships, a corporation might spin off certain environmentally related assets into a new partnership entity that could then sell units in the partnership to investors seeking a passive income stream.

Research and Development

Similarly, research and development (R&D) limited partnerships are also available as instruments for financing corporate environmental research and development. These partnerships would benefit by claiming tax deductions and sharing in the revenues, if any, resulting from enhanced production, asset utilization, or lifetime extension.

There is also an increasing need to link EHS excellence with financial success and to provide appropriate comparative data to the financial markets. This recognizes that there is a growing awareness of the financial implication of EHS activities within the investment community, and companies must ascertain and respect the financial industry's perspective. The financial sector may be comprised of a variety of constituent types—portfolio managers, corporate bankers, lenders, sell-side analysts, insurers, retail investors, institutional investors, investment bankers, and buy-side analysis—but all have common perspectives when it comes to assessing the financial implications of sound environmental management.

Environmental management attention-getters to the financial sector include EMS; training, auditing, and verification; compliance history; spill or incident history; health and safety performance; efficiency (energy, water, raw materials); and EHS cost trends. However, recognize that investors are also seeing eco-efficiency emerging as a proxy for competitive positioning. To that end, global reporting standards are increasing, emerging as basis for risk-adjusted pricing.

A primary motivation is to minimize losses from loan defaults, but there are also concerns regarding customer viability and lender liability. EHS credit risk may include the nature and extent of liabilities, current and future cap expenditures and operating costs, and the effectiveness of risk controls.

From the insurance industry perspective, the key motivation is to control losses from claims, and to that end there is concern as to the customer's ability to control risks. The EHS role in all this is to establish the company's EMS as a quality risk-control proxy to assuage insurance industry concerns.

From the investment community's perspective, the primary motivation is upside gains. However, this is often outweighed by concerns with regard to a loss of share value, revenues, and earnings as a result of unanticipated environmental loss events.

Historically, many investors have viewed environmental management as a liability, not an opportunity; as a moral and ethical issue, not financial; and from a financial standpoint, sound environmental management may not be a material issue from an upside standpoint. In that sense, environmental management has often been seen in the past by the investor community as a stand-alone entity that is not strategic to the company. However, there is a dichotomy here because many in the investor community are starting to believe that "non-financial" issues—including EHS—will become more important to the bottom line over the next five to ten years, and the same investors will pay a premium for companies whose EHS strategies lead to a competitive advantage.

Thus, new "environmentally sensitive" market trends are emerging that may have a profound impact on the sustainability of shareholder values:

- New market forces requiring environmental and social responsibility performance
- Growing interest in environmental and social issues of capital providers: banks and equity markets
- Growing needs to manage companies in a holistic way, allowing the senior managers to differentiate from sector and industry averages and to improve the corporations' competitiveness and sustainability, thus generating sustainable shareholder value

Environmental risk is financial risk. The following lists a few specific ways in which environmental risk translates into financial risk in a way that impacts and grabs the attention of the investor community:

- Balance sheet risk (historic liabilities, impairment of real property values, underwriting losses)
- Market risk (corporate reputation and image, reduced customer acceptance, potential loss of "social license to operate")
- Operating risk (emissions and discharge risk, product liability risk, required process changes)
- Capital cost risk (pollution control expenditures, product redesign costs)
- Transaction risk (potential cost of time, money, and delayed or canceled acquisitions or divestitures)
- Business sustainability risk (potential competitive risk from lack of efficiency or sustainability in energy, materials, and resource use)

Furthermore, eco-efficiency drives investment out-performance: superior eco-efficiency is a powerful tool for portfolio risk mitigation. Also, eco-efficiency is a robust leading indicator for sustainable earnings quality and shareholder value-creation going forward. There is a hidden value potential not currently captured by traditional Wall Street analytics; therefore, an information arbitrage opportunity exists because the future out-performance gap will be even greater.

A "green wall" has developed in corporate and investor circles regarding environmental management issues and, as a result, a significant gap has emerged. Why the green wall? A lack of understanding of environmental management's relevancy and language play a part, as well as company attitudes and a lack of understanding in the investment community. More and more, the gap can be bridged by establishing "best of class" investment and EHS value analytical tools.

Proactive environmental management calls for raising the consciousness of the audience via attention-getters such as:

- Trends
  - Consumer and public opinion, regulation
- Competitive curve shifts
  - Products, processes, strategy
- Performance impact
  - New business and market opportunities
  - Improved efficiencies and margins
  - Ability to meet growth and earnings targets
  - Consistency and predictability of earnings

Proactive environmental management also calls for engaging the financial community by taking the lead; making relevant contributions to the science; being consistent in public posturing; avoiding "greenwash;" establishing and using benchmarks vigorously to improve performance; and sending clear, positive signals regarding environmental stewardship goals.

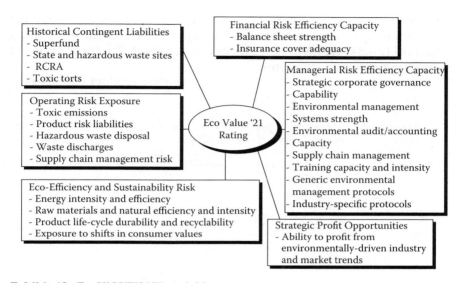

Exhibit 42. EcoVALUE'21™ variables.

Increasingly, the investment community is turning to independent eco-value systems to help them measure and understand these issues. A good example is the EcoVALUE'21™ model. EcoVALUE'21 is used as a tool to:

- Evaluate the environmental performance of companies within their sector
- Optimize a company's EHS activities to maximize shareholder value and financial performance
- Enhance long-term investment portfolios, such as corporate pension funds, to minimize environmental risk and maximize returns
- Identify hidden shareholder value and unrecognized differentials in eco-efficiency
- Improve due diligence on corporate acquisition

EcoVALUE'21 was established by Innovest strategic value advisors, a leading-edge international investment advisory firm founded in 1995 by Dr. Matthew Kiernan, the former director of the Business Council for Sustainable Development. Eco-VALUE'21™ is based on the evidence that eco-effectiveness is a proxy for, and predictor of, superior corporate management that generates superior financial performance and shareholder value. The EcoVALUE'21 analyzes over 60 key variables, using over 20 data sources. Exhibit 42 identifies some of the key variables.

The key variables are then aggregated and summarized in a scoring matrix. The raw scores are weighted using Innovest's proprietary investment algorithms and a final score is generated. The score is based on the company's environmental performance relative to its competitive set and are then converted to alphabetical ratings similar to the familiar ratings on corporate bonds. While still in the development stage, these types of indices are the wave of the future.

References

ASTM. "Standard Practice forEnvironmental Site Assessments: Phase I Environmental Site Assessment Process." (Philadelphia, PA: ASTM, 1993.)

Kiernan, Matthew J.: Eco-Value, Sustainability, and Shareholder Value: Driving Environmental Performance to the Bottom Line. *Environmental Quality Management*, 10:4; 1–12; 2001.

Chapter 10
Benchmark Survey

The keys to benchmarking success are to:

- Tie benchmarking to goal-setting
- Target best industry vs. average or composite
- Be selective in choosing benchmarks—keep asking the "so what" question
- Take management actions as a result of benchmarking data

Benchmarking should combine two dimensions to provide a truly comprehensive approach to assessing your environmental management (Exhibit 43). The benchmarking analysis should compare your practices with the "best," using both quantitative and qualitative analysis (Exhibit 44).

Exhibit 45 provides another way of looking at benchmarking. The internal view establishes the performance baselines and measures the impact of process improvement and other important initiatives. But the external view allows the company to determine best practices and, in turn, smoothly plan to adopt best practices. A benefit of benchmarking is that it is a basis for course corrections and other management decisions.

Like the other lines of inquiry in the environmental management assessment process, benchmarking depends on a comprehensive understanding of three key program elements: management effectiveness, technical understanding, and regulatory issues.

There are six key factors that contribute to successful benchmarking:

- Formulating a clear purpose of the benchmarking study
- Assessing relevant industry and trade association data with benchmark targets
- Involving the company in data selection within and perhaps in setting up the supporting review sheets
- Offering participating companies something tangible and valuable in return for their help (e.g., an evaluation of their operation, a shared list of best practices, or an early look at the report before it becomes generally available to their industry)
- Using personnel who are highly familiar with the industry and the issues being addressed in the benchmarking study
- Developing clear and concise questions that are focused on the specific issue at hand

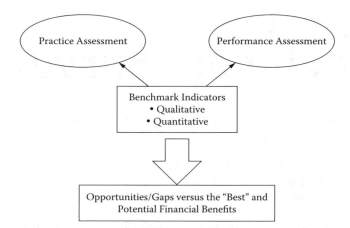

Exhibit 43. Benchmarking environmental practice and performance.

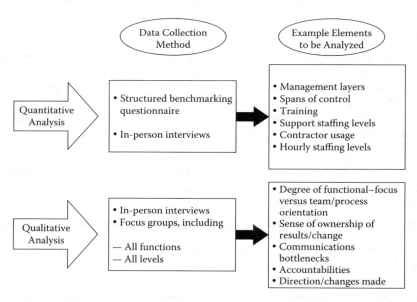

Exhibit 44. Example of process for benchmarking environmental management structure and effectiveness.

In addition to technical expertise, the benchmarkers must be intimately familiar with the statutory mandates governing today's environmental management efforts. Familiarity with the regulatory climate, coupled with extensive field experience, allows the evaluator to conduct technical benchmarking studies from both a performance and regulatory perspective. This capability is directly responsive to increasing public

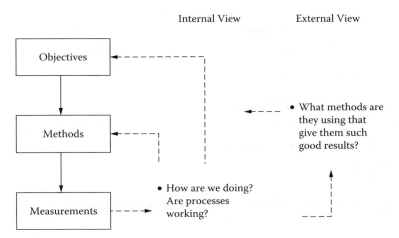

Exhibit 45. Difference between internal and external views gained from benchmarking.

demands that environmental policies function successfully under the broadest set of possible scenarios.

The benchmarking approach may vary from more management-oriented studies to more technical-oriented studies. Both types of benchmarking activities should include on-site interviews and document reviews. Brief visits to generation facilities are conducted to verify that environmental programs and procedures are being implemented and data are being accurately developed.

Environmental Management-Based Benchmarking Survey Approach

Environmental management assessments call for a more management-oriented approach along the lines of the 14 key functional criteria discussed previously and represented again in Exhibit 46. A technical-oriented approach may be added to supplement, as necessary. Results can be measured based on a Stages of Excellence Model (see Exhibit 47) to facilitate the easy understanding and communication of an organization's environmental goals, current position, and opportunities for improvement.

The environmental management benchmark survey should provide a comparative analysis of environmental policies, environmental procedures, staffing and organizational setup, environmental compliance, and environmental release data. These types of environmental management surveys may also be tailored to reflect specific industry concerns and regional or national issues, as determined by client needs.

The goal of the management-oriented benchmark analysis is to assess whether a company compares favorably with industry organization-wise

Senior Management Commitment	Operational Systems
Corporate Environmental Policy	Risk Assessment
External Communication	Audit Implementation
Strategic Environmental Plan	Issue Management
Organization and Staffing	Waste Minimization
Internal Communications	Environmental MIS
Internal Integration	Financial Tracking

Exhibit 46. Criteria for environmental management success.

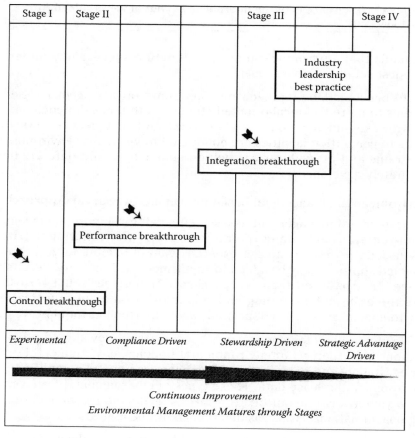

Exhibit 47. Stages of the environmental management success model.

```
Air emissions

RCRA hazardous waste generation

Combustion waste

Spills

Toxic release inventory (TRI) data

Notice of violations (NOVs)

NPDES permit exceedances

Opacity emission limit exceedances

Environmental expenditures
```

Exhibit 48. Performance parameter examples.

and compliance-wise—especially considering its size—and how further advancement can be achieved through the implementation of emerging best management practices.

Technical-Based Benchmarking Survey Approach

The approach to the supplemental technical-based benchmarking study is as follows:

- Establish and define what type of performance parameters are to be obtained.
- Develop a questionnaire that provides a framework for attaining the performance parameters identified in the previous step.
- Establish display plots that characterize the specific performance parameter for a combination of performance parameters that are to be evaluated.

Exhibit 48 provides a sampling of various performance parameters that may be applicable to environmental control for industrial facilities.

To generate the needed information, a questionnaire must be designed to collect specific data of interest. It is important that the participating companies are assured anonymity in the reporting process to ensure full cooperation and accuracy. Confidentiality can be accomplished by assigning each participating company a unique numeric code in the formal reporting process.

As shown in Exhibit 49, the questionnaire is typically organized with the data request in each row and descriptive information about the data in columns. The first two columns provide the category description and unit of measurement for the requested data. These columns are completed

Data Category Description	Unit of Measure	Data Collected	Notes
(1) All company operations			
(2) Specific equipment			
(3) Qualitative data on environmental management			
(4) Data packaging			

Exhibit 49. Sample format of a benchmarking questionnaire layout.

prior to a site visit. The third column ("Data") is where the requested numeric information is inserted. The final column ("Notes") is where the reviewers can add information to further clarify and explain special circumstances (such as an indication that units were shut down for x months during the calendar year, and so on).

The questionnaire can typically be divided into four sections. The first section is designed to collect numeric data for all company operations. For example, for a utility study this section would include the operation of conventional steam units as well as combustion turbines.

The second section collects numeric data specific to company-operated equipment of interest. Using the utility example, combustion turbines can be treated separately in this section, and in so doing compare the performance of combustion turbines independently from their other units.

The third section is typically designed to ask qualitative questions about the environmental management of the companies involved in this program. Again, as an incentive to participate we would recommend offering each participant a confidential qualitative profile of their performance against the other companies participating in the benchmarking program. Confidentiality would be maintained becausee companies would only be identified by a unique numeric code and each company is informed of only their specific numeric code. Thus, complications of joint information can be presented while still preserving confidentiality.

The last section of the questionnaire typically identifies the type of diskette and version of software on which we would like to receive the raw data provided by all participating companies. The data set should not reveal the identity of the companies providing specific data.

Successfully completing questionnaires requires the questioner to pay special attention to the definitions of the variables. In particular, there should

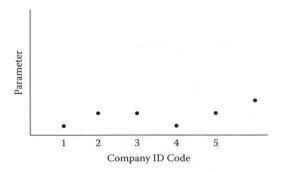

Exhibit 50. Sample of a display plot format.

be an emphasis on defining how to deal without company-operated units. Questionnaires typically focus on the units that are operated by a specific company. For those situations where a company reports figures for units that the company operates but jointly owns, the questionnaire could allow for reflecting only the percentage of the variables "owned" by their company.

For cases where a company is unable to provide the requested numeric value, checking one of the two boxes below the data item should indicate the reason. The first box ("Not Applicable") should be used to indicate that the requested data item does not exist at the company. The second box ("Not Reported") indicates that the data were not readily available or that the company has decided not to provide the information for the survey. For some data items, a firm may not be able to obtain a comparable measurement, or the effort required to collect the information more than offsets the benefits to the participating company. The "Not Applicable" versus the "Not Reported" is an important distinction. However, it should be emphasized to the participating company that the questionnaire will be unable to calculate those performance parameters for the company that are dependent upon data items that have been left blank. For some measurements, the correct answer for a company may be "zero." This value should be recorded in the data column as the number zero.

As depicted by Exhibit 50, display plots for each of the established calculated parameters for all participating companies that submitted their data are developed and provide an indication of the relative performance for each company. The title for each plot unit will typically contain the parameter number and the formula (if the plot is based on a combination of parameters). The vertical axis shows the value of the measure, the measurement units, and a scalar. The horizontal axis identifies each company by its confidential identification (ID) number.

There is an increasing database being developed to capture industry environmental data in a meaningful way that can be incorporated into

Document number	Industry
EPA/310-R-95-001	Dry cleaning industry
EPA/310-R-95-002	Electronics and computer industry*
EPA/310-R-95-003	Wood furniture and fixtures
EPA/310-R-95-004	Inorganic chemical industry*
EPA/310-R-95-005	Iron and steel industry
EPA/310-R-95-006	Lumber and wood products industry
EPA/310-R-95-007	Fabricated metal products industry*
EPA/310-R-95-008	Metal mining industry
EPA/310-R-95-009	Motor vehicle assembly
EPA/310-R-95-010	Nonferrous metals industry
EPA/310-R-95-011	Non-fuel, non-metal mining industry
EPA/310-R-95-012	Organic chemical industry*
EPA/310-R-95-013	Petroleum refining industry
EPA/310-R-95-014	Printing industry
EPA/310-R-95-015	Pulp and paper industry
EPA/310-R-95-016	Rubber and plastic industry
EPA/310-R-95-017	Stone, clay, glass, and concrete industry
EPA/310-R-95-018	Transportation equipment cleaning industry
EPA/310-R-97-001	Air transportation industry
EPA/310-R-97-002	Ground transportation industry

Exhibit 51. List of the EPA's Office of Compliance Industry Sector Notebooks.

benchmarking analysis. One potential resource is the EPA's Office of Compliance Industry Sector Notebook. A list of these reports is presented in Exhibit 51, and samples of information available are presented in Exhibit 52, Exhibit 53, and Exhibit 54.

Other sources of information are trade associations. The American Petroleum Institute (API) is a good example. API is the largest trade group for the petroleum refining industry, with the largest membership and budget. API represents the major oil companies and independent oil producers, refiners, marketers, and transporters of crude oil, lubricating oil,

EPA/310-R-97-003	Water transportation industry
EPA/310-R-97-004	Metal casing industry
EPA/310-R-97-005	Pharmaceuticals industry
EPA/310-R-97-006	Plastic resin and manmade fiber
EPA/310-R-97-007	Fossil fuel electric power
EPA/310-R-97-008	Shipbuilding and repair industry
EPA/310-R-97-009	Textile industry
EPA/310-R-97-010	Sector notebook data refresh—1997 Seth Heminway
EPA/310-R-98-001	Aerospace industry
EPA/310-R-98-002	Agricultural chemical, pesticide
EPA/310-R-98-003	Agricultural crop production
EPA/310-R-98-004	Agricultural livestock production
EPA/310-R-98-005	Oil and gas exploration
EPA/310-R-98-008	Local government operations

*Spanish translations available

Exhibit 51 (continued). List of the EPA's Office of Compliance Industry Sector Notebooks.

gasoline, and natural gas. API conducts and promotes research in the petroleum industry and collects data and publishes statistical reports on oil production and refining. Numerous manuals, booklets, and other materials are published on petroleum refining to assist members in environmental compliance. The Electric Power Research Institute (EPRI) is a comparable example for the utility industry.

Finally, as noted previously private code organizations are a good source of industry data and research as well:

- CERES
- GEMI
- ISO 14000
- The United Nations Environmental Program (UNEP) Finance Initiatives
- Canada's World Business Council for Sustainable Development
- The International Institute for Sustainable Development
- The Environmental Banking Association
- The World Resources Institute

Industry Sector	CO	NO$_2$	PM$_{10}$	PT	SO$_2$	VOC
Metal mining	5,391	28,583	39,359	140,052	84,222	1,283
Nonmetal mining	4,525	28,804	59,305	167,948	24,129	1,736
Lumber and wood production	123,756	42,658	14,135	63,761	9,149	41,423
Furniture and fixtures	2,069	2,981	2,165	3,178	1,606	59,426
Pulp and paper	624,291	394,448	35,579	113,571	541,002	96,875
Printing	8,463	4,915	399	1,031	1,728	101,537
Inorganic chemicals	166,147	103,575	4,107	39,062	182,189	52,091
Organic chemicals	146,947	236,826	26,493	44,860	132,459	201,888
Petroleum refining	419,311	380,641	18,787	36,877	648,155	369,058
Rubber and misc. plastics	2,090	11,914	2,407	5,355	29,364	140,741
Stone, clay and concrete	58,043	338,482	74,623	171,853	339,216	30,262
Iron and steel	1,518,642	138,985	42,368	83,017	238,268	82,292
Nonferrous metals	448,758	55,658	20,074	22,490	373,007	27,375
Fabricated metals	3,851	16,424	1,185	3,136	4,019	102,186
Computer and office equipment	24	0	0	0	0	0
Electronics and other electrical equipment and components	367	1,129	207	293	453	4,854
Motor vehicles, bodies, parts and accessories	35,303	23,725	2,406	12,853	25,462	101,275
Dry cleaning	101	179	3	28	152	7,310

Source: U.S. EPA Office of Air and Radiation, AIRS database, May 1995

Exhibit 52. 1993 Pollutant Releases (short tons/year), from the Petroleum Refining Industry Sector Notebook, September 1995.

References

ISO. "Environmental Management Systems—Specification with Guidance for Use." Reference number ISO 14001:1996(E). (West Conshohocken, PA: ASTM. 1996.)

U.S. EPA's Office of Compliance: Industry Sector Notebooks; EPA/310-R-95-001 through 95-018; EPA/310-R-97-001 through 97-010; and EPA/310-R-98-001 through 95-008, 2001.

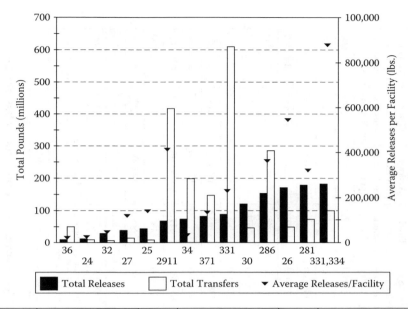

SIC Range	Industry Sector	SIC Range	Industry Sector	SIC Range	Industry Sector
36	Electronic equipment and components	2911	Petroleum refining	286	Organic chemical mfg.
34	Lumber and wood products	34	Fabricated metals	26	Pulp and paper
32	Stone, clay, and concrete	371	Motor vehicles, bodies, parts, and accessories	281	Inorganic chemical mfg.
27	Painting	331	Iron and steel	333, 334	Nonferrous metals
25	Wood furniture and fixtures	30	Rubber and misc. plastics		

Exhibit 53. Summary of 1993 Toxics Release Inventory (TRI) data: Releases and transfers by industry, from the Petroleum Refinery Industry Sector Notebooks, September 1995.

Industry sector	SIC range	# TRI facilities	1993 TRI releases		1993 TRI transfers		Total releases + transfers (million lbs.)	Average releases + transfers/ facility (lbs.)
			Total releases (million lbs.)	Average releases per facility (pounds)	Total transfers (million lbs.)	Average transfers per facility (pounds)		
Stone, clay & concrete	32	634	26.6	42,000	2.2	4,000	28.8	46,000
Lumber and wood products	24	491	8.4	17,000	3.5	7,000	11.9	24,000
Furniture & fixtures	25	313	42.2	135,000	4.2	13,000	46.4	148,000
Printing	27	318	36.5	115,000	10.2	32,000	46.7	147,000
Electronic equip. & components	36	406	6.7	17,000	47.1	116,000	53.7	133,000
Rubber & misc. plastics	30	1,579	118.4	75,000	45	29,000	163.4	104,000
Motor vehicles, bodies parts, & accessories	371	609	79.3	130,000	145.5	239,000	224.8	369,000
Pulp & paper	2611-2631	309	169.7	549,000	48.4	157,000	218.1	706,000
Inorganic chem. mfg.	281	555	179.6	324,000	70	126,000	249.7	450,000
Petroleum refining	**2911**	**159**	**64.3**	**404,000**	**417.5**	**2,625,000**	**481.9**	**3,088,000**
Fabricated metals	34	2,363	72	30,000	195.7	83,000	267.7	123,000
Iron & steel	331	381	85.8	225,000	609.5	1,600,000	695.3	1,825,000
Nonferrous metals	333, 334	208	182.5	877,000	98.2	472,000	280.7	1,349,000
Organic chemical mfg.	286	417	151.6	364,000	286.7	688,000	438.4	1,052,000
Metal mining	10		Industry sector not subject to TRI reporting					
Nonmetal mining	14		Industry sector not subject to TRI reporting					
Dry cleaning	7216		Industry sector not subject to TRI reporting					

Exhibit 54. 1993 Toxics Release Inventory (TRI) data for selected industries (Source: Petroleum Refinery Industry Sector Notebooks, September 1995).

Chapter 11
External Survey

As discussed in Chapter 2, the external survey can be comprised of one to three elements. The external scan is the central, core element and can be supplemented as deemed necessary by an independent technology scan and a national resource damage assessment scan.

External Scan

The external scan encompasses confidential face-to-face and, in some cases, telephone interviews with a range of stakeholders from outside the company. Typical participants in the external scan include:

- The political community;
- The regulatory community; and
- Public advocacy groups.

Exhibit 55 presents a representative list of the types of organizations that may typically be included in the external scan. It is critical that the information collected be confidential and based on interviews conducted by independent third parties.

Assessing Global Impacts—Sustainable Development

For those organizations that have global operations and commitments, a more expanded external scan is necessary that assesses the global impacts of company environmental policies and the degree of consistency and integration, given the potentially disparate regulatory situations and the long-term strategic implications of the company's environmental policies and posture within the relevant world community.

Almost 50% of the *Fortune* global top 250 companies are issuing environmental, social, or sustainability reports. Increasingly, organizations are seeing this as part of a legitimate process to maintain a "social license to operate" in a global forum that is increasingly sensitive to sustainable development issues.

Sustainable development is about balancing the economic, social, and environmental issues over the short- and long-term to ensure a viable business climate. As such, sustainable development is inherently in sync with good business because "good business" is all about balancing issues.

- Governor's Office
- State Environmental Protection Agency
- Public Service Commission
- Industrial Associations
- U.S. Bureau of Land Management
- U.S. Fish and Wildlife Service
- U.S. Forest Service
- U.S. EPA Regional Office
- County Environmental Officials
- City Environmental Officials

Exhibit 55. Potential external scan participants.

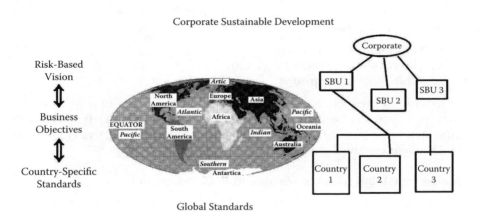

Corporate Sustainable Development

Exhibit 56. Assessing global impacts.

Global forums and matrices are already starting to emerge through the development of the Dow Jones Sustainability Index, the FTSE4 Good Index, and the UK's Association of British Insurers Socially Responsible Investment Guidelines. These initiatives provide a mechanism for aligning sustainable development initiatives with shareholder interests and are a response to the various drivers spurring sustainable development (see Exhibit 57).

Ironically, sustainable development and related greenhouse gas emission reduction initiatives have less to do with environmental public relations and more to do with a drive to increase long-term operational and resource efficiencies with concomitant positive impacts on the corporation's bottom line. Many organizations are finding that efficiency upgrades and sustainable

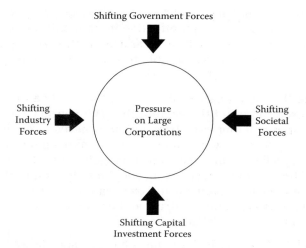

Exhibit 57. Forces driving sustainable development.

development initiatives generate paybacks on the order of five years, after which savings and enhanced opportunities turn to sheer profit.

Strategic management of environmentally related sustainable development performance is particularly difficult to assess and manage on a global basis. Differences in regulatory climates and approaches severely impact the development and implementation of uniform management standards that can be applied worldwide. Furthermore, applying emergency response programs and related community communications on a consistent basis globally calls for intensive and ongoing training of key personnel and environmental management system development that far exceeds historical, more episodic risk management efforts. In short, new global forces have and will continue to emerge that require increased environmental responsibility (e.g., Kyoto). Taking advantage of short-term differentials in national environmental policies may provide short-term advantage but they open firms up to the long-term risks with both operational and political repercussions.

Recognize too that recent developments in national air emission and water discharge trading schemes and wetland/wildlife preserve banking have provided industry with an opportunity to translate strong sustainable development policies into tradable assets. This is particularly relevant from a global perspective, where such practices have been established for a longer period of time.

Whereas sustainable development may have a negative impact on operational asset utilization, more efficient management of the issue in comparison to the competition may also provide a significant asset edge. In some instances, organizations have found that their facilities and operations may have strong secondary market potentials responding to mainstream environmental dilemmas. Also, effective implementation of innovative environmental technology can be used as a market weapon against laggard competitors who are more vulnerable to evolving regulatory policy and cost implications.

Recognize that sustainable development's roots can be traced to a gradually increasing trend throughout the 20th century for government involvement in regulating industrial and natural resource development. The 1960s and 1970s saw an emergence of environmental and social concerns being reflected in legislation and regulations. The latter could be seen as a statement of society's values and a reflection of the growth in power of both large multinational and transnational organizations and the concomitant growth in public mistrust and demand for direct involvement in development decision-making processes.

These societal forces reflect not only changing societal values and community expectations but also an emerging emphasis on indigenous rights. In turn, these forces have inspired shifting governmental priorities—most notably in the environmental and community health and economic development. However, not only shifting societal and government forces have shaped the need for sustainable development but also shifting industry and capital investment forces.

The capital investment community is not only subject to the influence of worldwide economic and financial pressures but also to pressures from aggressive stakeholder and advocacy investment communities who are demanding an evolution in the investment community's role. Financial institutions are increasingly seen as tribunals for airing development disputes, and these disputes are not just impacting public sector investors but also private investors. The capital investment is increasingly challenged to accurately gauge the inherent bottom-line financial risks of adverse societal reactions. This may be seen in both an evolving "social license to operate" spirit as well as the potential risks to corporate brand reputation due to preventable destructive incidents.

Industry itself is not only subject to societal, government, and economic and financial forces but also to internal change forces. The latter may be seen as industry restructuring occurs, global materials and market competition intensifies, new technologies and business models emerge, and competition for human resources not only intensifies but is redefined by societal expectations and industry's need for an increasingly educated work force.

To accurately gauge this challenging business environment, organizations must engage in a much broader, more thoughtful strategic planning exercise that not only addresses financial planning but also emphasizes scenario

Exhibit 58. Scenario for building the business environment.

development, assessment, and counter-planning to safeguard capital asset investment and revenue streams in the long-term. The latter must not only reflect the influence of economic performance but also the myriad societal values that may emerge and influence the corporation's bottom line.

Briefly, the optimistic business scenario is obviously the goal. It is an environment where economic growth is high and the social forces are both peaceful and unifying. In short, economic forces and social forces are in sync. It is a goal but rarely reality; ironically, it is often the basis for corporate planning. More realistically, corporations find themselves in challenging environments where difficult times are being met by a drive for innovation and newfound societal compacts that nurture respect and cooperation.

The pillaging and downbeat environments are the most problematic but in different ways. In the downbeat environment, depressed economic conditions coincide with tumultuous social conditions that can result in a dangerous downward spiraling that can have devastating financial impacts. The pillaging scenario is characterized by an unhealthy distancing in financial health between societal segments and is vulnerable to strong potential political turmoil that could result in swift and calamitous changes in economic and financial conditions.

The patterns of movement between scenarios are reflected as clockwise and counter-clockwise and if unchecked, the likelihood of movement is reflected by the intensity of the arrows (shown in Exhibit 58).

Project Life Cycle Analysis

Sustainable development translates into the need to measure the value created by organizations relative to a "triple bottom line" of economic prosperity, environmental feasibility, and social responsibility. This recognizes that the very nature of development activity necessitates a firm, stable foothold in the community.

This calls for respect for the roles of individuals inside and outside the company, as well as recognition of the importance of cultural diversity and environmental stewardship. The latter should be addressed not merely as a constraint but as an opportunity for competitive advantage. Sustainable development complements other stewardship concepts such as project life cycle management. The latter is an effort to look at the entire impact of proposed projects, providing process actions for the triple-bottom-line prospect of sustainable development. Typically, the life cycle may be looked at in four phases—evaluation, implementation, operation, and closure—with variations in specifics from industry to industry.

The *evaluation* phase is the starting point in project/product/process management. From a material resource perspective, this might be the exploration phase, or from a manufacturing/retail perspective, new plant sighting in product/process studies. The objective is to be quick and efficient yet thorough, keeping the evaluations as closely vested as possible to avoid proactive competitor response and unjustified public concerns and expectations. Because of the inherent secretive nature of this phase, social and environmental implications are often minimized or ignored. However, this is just the area where a strong, independent yet confidential review of such issues can be significant to head off unwarranted development actions and the possible risk of subsequent costly snags, delays, or failures.

The initial *implementation* phase is a critical juncture. Often, it involves an intense set of activities that can lead to severe problems if not carefully managed. Typically, this is the project/product/process design and con-struct phase. From a project construction standpoint, this phase often involves the dislocating impact of a much larger, short-term workforce and community infrastructure description, and the stress implications are enormous if not carefully planned. From a product/process development standpoint, decisions made at this critical phase will be the foundation for future social and environmental impacts. Product redesign and process redesign costs to accommodate tardily recognized social and environmen-tal implications can be enormous in both costs and reputation. Thus, from all these perspectives (project/product/process) effective systems for ensuring good decision-making and practices are essential. Note that this phase marks the time that formal approaches are sought and the first opportunity to aggressively engage the public.

Ironically, the *operation* phase often receives the lion's share of attention despite the fact that many critical decisions have preceded it. Still, in the general public's mind it is the operating phase that conjures up the most vivid images. Typically, economic, social, cultural, and environmental impacts are well understood but they may be addressed in an uneven fashion.

The *closure* phase may be either final or temporary. The various aspects of closure may occur due to changing product or process lines, changes in pricing, accident, disaster, labor strife, or government nationalization actions. Closure rarely receives the consideration and planning that it warrants. Human nature possibly inhibits those who are involved with creation to adequately assess burial rites. This is another area where some independent, "devil's advocate" perspective might be helpful.

In essence, corporate sustainability is about creating long-term shareholder value by embracing—not shrinking from—opportunities by actively managing the economic, environmental, and social risks as well as the benefits posed by development.

In summary, the questions that must be addressed during the four-phase project life cycle review include:

- Are the cost/benefit risks equally distributed?
- What are the implementation and operational tradeoffs that must be considered?
- What are the market tradeoffs that must be considered?
- How does this project fit within the whole system?
- What are the uncertainties, and what precautionary measures and management adaptations may be necessary?
- What is the cumulative impact of the project, and how does the project fit to the current business landscape?
- What steps must be taken to arrive at synthesis and consensus with the disparate interested parties?

Activist Group Alliances

It is critical to employ conservation science as a foundation of any environmental policy; unfortunately, this is often seen as a rather innovative approach. The approach also calls for stressing innovations and selection through discussions and evaluations with the entire supply chain with the stakeholders' reaction in mind. This involves more than just looking at "end-of-stack" environmental consequences. To develop policies that steer the demand for more environmentally sound products and reflect the company's values while achieving measurable, quantifiable results over time, reports should be responsible, transparent, and accountable.

"Responsible" is essentially being very specific. The issues are not continuously morphing; they are very defined. Stability increases certainty, which facilitates transparency, which in turn enables the corporation to be accountable. It is employing environmental strategy that is responsible, developing policies that are transparent in an annual mission statement with goals and an annual mission statement that ensures that you are accountable.

CORPORATE ENVIRONMENTAL MANAGEMENT

The influence of the consumer can have a powerful effect on environmental management, and often this involves educating the customer. Environmentally sound products do not imply low quality. Environmental management programs are value-based efforts but they must be market-driven as well. The true objective is to derive economic benefit from improved environmental outcomes. To do so, environmental management programs must access the business model and environmental performance into the business model. To achieve this, environmental management must be fully integrated with the stakeholder community as well as with the entire organization. For a start, environmental management must meet with as many people of as many functions within the organization as possible and establish how environmental performance can add value to that business unit. Likewise, a strong program to secure the buy-in from your key suppliers is essential. This includes assessing what type of leverage the corporation has to achieve its goals with suppliers. Finally, a strong external communication program must be in play to communicate these efforts to the stakeholder community and gain critical feedback.

A successful foundation of business conservation science management is a collaborative approach with suppliers, NGOs, and many other stakeholders, including our sales teams. Again, it must be market-driven yet value-based. The economic leverage with suppliers—although it is there—must be secondary to the overall buy-in on market strategy impact. By focusing on the process-based sources of value where you can actually save money or reduce costs, environmental management can quickly win over colleagues. Product differentiation can be achieved by increasing the product receipts value and by meeting the environmental attributes customers need. In the end, it results in winning new business rather than a cost reduction.

However, significant cost reductions are available. For example, by just shifting delivery trucks to ultra-low-emissions vehicles, firms are finding that it can reduce greenhouse gas emissions by 40% but also reduce fuel costs by 25% and increase the productivity of the workforce. Unfortunately, many companies have an environmental policy but no strategy or approach to implementation. In these cases, their environmental reports provide the stakeholder with an overview of the corporate policy without providing the infrastructure put in place, the initiatives and programs implemented, and the resulting environmental outcomes that can be verified by an independent party. Recognize that whereas campaigning environmental groups (i.e., activist groups) have value in escalating issues into the general public awareness and into executive offices, their utility tends to fade because once the issues are raised, the important thing to focus on is solutions. Still, corporate environmental management must be realistic in their discussions with activist groups. Activist groups will basically take credit for initiatives and use that to raise money for their next campaign. The goal is to not be a poster child for their funding campaigns and, when possible, establish production allies in the activist group community.

For retailers, environmental policies offer firms rewards around recycling and pollution reduction, developing awareness of environmental issues and markets for environmentally preferable products, and in some cases forest and biodiversity conservation. A firm has direct control over the first two issues but with the third issue, a firm relies on supply chain partners. Activist group alliances can impart some influence on economic and social stakeholders by raising awareness and changing the game in how issues are approached, and generate support for a solution-focused, value-based, market-driven approach. Supply partners must be focused on solutions and committed to environmental issues rather than managing the risk associated with these issues.

Recognize that there will be competition among the various environmental certification schemes, and an important part of environmental management will be to assess and choose those certifications that are important to them and look past the certificate process to truly understand the interplay of environmental issues to the company's value.

Independent Technology Scan

The last element of the external scan process is the independent technology scan. This scan is more related to heavy industry, extraction industry, and medium-impact industries versus retailers. The independent technology scan is an effort to upgrade technologies and processes to do it faster, cheaper, and with less waste and environmental consequences.

The independent technology scan element involves surveying a wide range of vendor, supplier, and remedial action consultants to provide independent feedback on costs, schedules, and productivity issues that may impact the company's environmental management program. As part of the technology scan element, independent data on technology performance pertaining to given technical, cost, schedule, and regulatory criteria should be reviewed, and just as importantly, evaluated based on the "success criteria" identified in Exhibit 59.

The technology scan element identifies engineering and technology development opportunities, technology development opportunities being defined as requiring several years of significant R&D into technology applications that do not exist, whereas engineering developments are adaptations or modifications of currently available equipment systems and technologies. Any successful environmental restoration program must be prepared to track and take advantage of both opportunities.

- Enhances process safety
- Increases productivity
- Decreases cost
- Provides a cleaner end product
- Reduces waste volume
- Speeds up decision-making and screening criteria

Exhibit 59. Environmental management audit technology success criteria.

Chapter 12
Natural Resource Damage Assessment— Proactive Strategies

The thrust of this activity is how to avoid being a target of a *natural resource damage* (NRD) claim, or if you cannot avoid becoming the target of a claim, at least do the best job you can to prepare and position yourself effectively. The assessment entails finding out (through knowledgeable third parties) whether any trustee agency has initiated an NRD review and (if so) what its review criteria and priorities may be, identifying others in similar circumstances, and determining the basis for and scope of the claim. Exhibit 60 presents the framework for NRD claims. For the trustees to prevail in an NRD claim, all three of these elements must be demonstrated.

Essentially, *natural resource damage assessments* (NRDA) address that area where Superfund remediation leaves off. Why are NRD claims of concern? NRD claims can have devastating economic impacts because "cleanup" liability is typically based on full restoration without regard to risks, costs, or benefits. The stakes are particularly high where potential impaired reproduction of numerous fish, bird, and mammal species may have occurred. The definition of protected "natural resources" is extremely broad; for instance, compensable injuries could include the public's lost "psychological" enjoyment of resources for which they are unlikely to ever use. Companies associated with long-term aquatic sediment contamination may be most vulnerable to NRD claims. Because of the high financial stakes involved with NRDAs, it behooves an organization to develop proactive strategies to identify, shore up, and resolve vulnerabilities to claims.

The frustrating aspect to NRD issues is that the targets are unpredictable. Whereas only designated "trustee" agencies can initiate NRD claims, Federal trustees are becoming increasingly but sporadically active, and some states (e.g., California, New York, New Jersey, Ohio, Michigan, Colorado) are

Exhibit 60. Framework for natural resource damage (NRD) claims.

becoming more active under both federal and parallel state law. In addition, Native American tribes are beginning to get involved and, in some instances, are the dominant interest parties. It should be noted that NRD targets are not limited to designated Superfund sites. Damage payment by a single party at an individual NRD site can approach or exceed $100M.

Exhibit 61 depicts the NRDA process that a trustee must follow. Any trustee can trigger the process and the trustee receives rebuttable presumption, e.g., the potentially responsible party (PRP) must prove to the court that this process is inappropriate for the types of damages being alleged. As seen in this figure, the NRDA process includes pre-assessment, development of a damage assessment plan, the performance of damage assessment studies for each injured resource, and post-assessment activities. The process results in many steps and can require dozens of separate studies, reports, and public comment periods unless an early settlement can be reached.

Exhibit 62 provides two prototypical examples of the impact of the NRDA expansion. The stakes are particularly high where potential contamination of very large areas and potential impaired reproduction of numerous fish, bird, and mammal species may have occurred.

Companies associated with long-term aquatic sediment contamination may be most vulnerable to NRD claims. Thus, it is critical for large-scale environmental restoration sites to take NRD potential into account during the baseline engineering study development.

Exhibit 63 defines the types of environmental restoration sites that should be concerned about NRD claims. It should be noted that most NRD cases to date have involved contamination of aquatic sediments by "notorious" toxic pollutants.

In effect, an organization can follow two strategies: either lie back and wait, or be proactive. The advantages of a proactive approach are that the advance planning keeps transaction costs down and reduces ultimate NRD

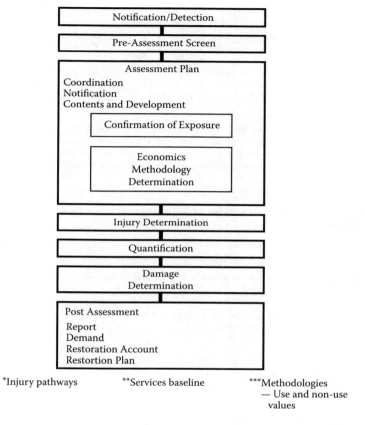

Exhibit 61. Natural resource damage (NRD) assessment process (as established by 43 CFR Part II).

liability exposure. The proactive strategy allows a site to identify potential contributors to future NRD actions and put programs in place to curtail the environmental restoration program's impact on the community as a whole and mitigate grounds for the NRD's action. It should also be noted that the recalcitrant who "lie back and wait" make more attractive targets.

If you cannot avoid becoming the target of an NRD claim, at least do the best job you can to prepare and position yourself effectively. Some of the concrete ways to avoid becoming a target are to find out (through a knowledgeable third party) whether a trustee agency has initiated an NRD review and (if not) what its review criteria and priorities are. If you are a Superfund PRP, be sure that potential NRD issues are addressed during the remediation phase and that appropriate legal releases are obtained. In particular, the site should consider cost-effective mitigation measures to minimize the impact of environmental restoration activities. In short, make sure you do

111

- The concept of non-use economic benefits is still evolving, but includes such components as bequest and experience values that add to a PRP's liability. The bequest value is the value associated with the knowledge that a resource is preserved for future generations. The existence value is the value associated with the knowledge that a resource is preserved, regardless of its use-related benefit to a person. For example, one may never visit the Everglades, but still attach a value to its existence in a healthy state (existence value) and to the knowledge that it will be around for one's children and grandchildren to enjoy (bequest value).

- The trustee can include any agency which "appertains to" or otherwise "controls" the resource, or any quasi-government organization with jurisdiction over some aspect of a resource. For example, Canada could conceivably file a case based on the loss of waterfowl that migrate to Canada. Likewise, a Native American tribe denied access to migratory salmon could initiate a case.

Exhibit 62. Prototypical examples of natural resource damage (NRD) claim expansion impact.

The types of environmental restortion sites that should worry about NRD claims:

- Are on a governmental hazardous substance list (e.g., CERCLIS, TRI database, NPL, CERCLA/SARA notification)

- Have been operating for 20+ years

- Have released large (over time) quantities of persistent, toxic, and bioaccumulative substances—especially if "notorious" (PCBs, DDT, dioxin, Pb, Hg, Cd, As)

- Are adjacent to "valuable" waterways, wetlands, waterfowl/wildlife refuges, endangered species habitat, tribal lands

- Are located in states with active NRD programs

- Do not have federal permits that fully and specifically authorize the hazardous substance releases in question

- Have contributed to extensive sediment contamination

> **It should be noted that most NRD cases to date have involved contamination of aquatic sediments by "notorious" toxic pollutants**

Exhibit 63. Types of environmental restoration sites.

not make the problem worse. In particular, characterization and remediation efforts should also include gathering evidence in support of potential NRD "defenses" (e.g., pre-1980 release, federally permitted release). Other proactive actions include beginning to assemble exculpatory records and identifying other potential NRD contributors.

The most effective way to reduce transaction costs and the size of settlements or awards in an NRD case is to keep the scope of the Damage Assessment Plan (DAP) and the Identification of Damaged Parties (IDP) within proper bounds. Exhibit 64 lays out ten technical defense tips to mitigate NRD claims. Focus early and often on the following likely "weak links" in the trustees' case.

If the site is a potential target of an NRD claim, a Site NRDA Model should be developed to assess whether the NRD issues are appropriately addressed during the remediation phase at the particular site through cost-effective mitigation measures.

As stated earlier, the concept of what constitutes a natural resource is subject to damages, and the nature of those damages has expanded. The PRP is now potentially liable for both residual damages and restoration (or replacement) costs unless the PRP can prove that the restoration costs are "grossly disproportionate" (e.g., *Ohio v. U.S. Department of the Interior*, 1989). Unfortunately, there are no clear definitions of "feasibility" or "grossly disproportionate," thus making it imperative that the potential PRP be proactive in defining these terms relative to his case.

The Site NRDA Model provides the basis for a site to effectively gather evidence in support of potential NRD "defenses" (e.g., pre-1980 release, federally permitted release). The successful site management will focus early on the areas where potential "unrealistic expectations" in the trustees' case may be. Exhibit 64 identifies ten areas of "unrealistic expectations" that the model looks for in an NRD claim. The list also provides a set of criteria to use to effectively reduce transaction costs and the potential size of settlements or awards in cases where claims may be valid but expectations are unrealistic. Again, the critical issue in the latter circumstance is to keep the scope of the DAP and the IDP within proper bounds to ensure that an effective cleanup and improvement is not delayed through legal "brain lock."

Different strategies are available for dealing with and mitigating NRD claims. A combination of approaches can be used to resolve NRD problems, but most of these strategies are associated with the timing or phasing of resource restoration or enhancement initiatives as part of cleanup operations. The following provides a discussion of alternate approaches to NRD issues at different stages of cleanup.

Early Recognition of Contamination Stage

Provisions of the revised NRDA rule expand the acceptable restoration activities to include restoring the resource, off-site restoration of a related service (e.g., creation of a wetland at an alterative location), off-site mitigation (e.g., wildlife habitat enhancement at a different location), and even the off-site protection of threatened resources (e.g., purchase of

1. Size of the area allegedly impacted as a result of the claimed releases. Size of the study area was part of the study area polluted/affected even prior to the release "baseline"? Keep the targeted area as small as possible relative to the overall size of the facility.

2. Quantifiable and compensable injuries. Eliminate any "injuries" that cannot be translated into quantifiable lost uses or "services." Even if you can quantify the number of common and abundant fish affected, compensable injury must be measured on an incremental basis. Incremental losses to an abundant resource may be small.

3. Components of the injury. The key here is to make sure you are not blamed for injuries due to other causes. Segment cause-and-effect relationships based on different contributory pollutants, different congeners, and other chemical, physical, and biological factors.

4. Uniqueness of the resource. Injury to rare and unique, rather than common and abundant, species. The key here is that claims based on injuries to common and abundant natural resources should be measured at the margin.

5. Moving "targets." Can injury incurred by migratory or highly mobile birds and mammals be potentially attributed to other geographically-specific release source(s)?

6. Evidence of actual harm. The practicality of the links between available lab and field data and evidence of biological harm (to individuals, a population, or the species). Under what circumstances is it permissible to extrapolate from one species to another or to select one species as an indicator or surrogate?

7. Status of population. Assess evidence of population impacts. Be skeptical of damage claims for species that are thriving in the wild.

8. Scope of damages. Assess the relevance of pre-release "baseline(s)," rough mass-balance calculations, nearby control reference areas, and applicable exclusions from liability.

9. "Per Se" injury. The applicability of "injury per se" criteria. The "injured" that are easiest to prove are those that are set forth as specific quantifiable criteria in DOI regulations. Examples are violations of water quality standards, levels in edible tissues in excess of FDA action levels, and levels in excess of those set by a state health agency in a directive to limit or ban consumption. However, recognize that contaminant levels in the surface microlayer cannot be directly compared to water quality standard/criterion levels. Also, note that exceeding a level set in a state fishing "advisory" is not the same as exceeding a limit contained in a consumption ban.

10. Feasibility of restoration. Recovered NRD monies can only be used for restoration. If restoration is not feasible, therefore, an NRD claim is not appropriate and resources should not be spent on assessment or quantification. Even where restoration is feasible, assessment is not appropriate where the cost will be "grossly disproportionate" to the value of the NRD claim.

Exhibit 64. Ten technical defense tips: Areas of potentially "unrealistic expectations" on which to focus.

some critical land for a new refuge or park). If adopted early, low-cost, off-site mitigation activities can greatly reduce a PRP's liabilities. For example, the placing of a relatively inexpensive salmon hatchery upstream or downstream of a potentially contaminated area near a salmon fishery can more than offset a much more expensive future NRD claim associated with possible losses to the fishery.

Agreement or Settlement Stage

Any official agreement (e.g., record of decision, consent decree, or settlement agreement) for site cleanup should attempt to include an explicit release from possible future NRD claims. Cooperation will be needed from the trustee agency or agencies, but this may be possible especially if proactive restoration or mitigation activities (such as those described previously) become part of an agreement. This is also an approach that should be sought in any renegotiation of existing agreements. In general, a PRP's leverage to receive a covenant not to sue is much stronger when it is part of a broader agreement rather than during litigation, where the NRD process gives the trustee many advantages.

Cleanup Stage

If a site is undergoing cleanup or remediation of contamination under CERCLA, an effort should be made to address cleanup and possible NRD problems simultaneously during all of the site investigation, planning, and implementation steps. By considering potential NRD problems during cleanup, this will minimize the exposure to future damage claims and provide the basis for covenant not to sue.

Pursuit of a PRP Claim Stage

At this stage, litigation or settlement alternatives loom. The company may wish to have a trustee show cause. Major tests must be demonstrated by a trustee to recover for resource damages. For example, as discussed in Exhibit 60, an NRDA claimant must demonstrate that:

1. Releases are above thresholds required to inflict an injury;
2. There is a casual link between the release and injury; and
3. Damages must be quantifiable and compensable.

If any one of these criteria cannot be met beyond a reasonable doubt, then there is no basis for proceeding with all the other aspects of an NRDA claim. Further, there are a number of general defenses that have been successful in reducing claims by 90% to 95%, and this represents an option for dealing with any NRD claim. However, this approach is not one that would necessarily enhance an already damaged public image of the company, and the latter should be strongly considered. Industry and the government federal facility complex are today paying a heavy price for yesterday's temporarily wise recalcitrance.

References

U.S. Department of Interior; 40CFR11.23 (f).
U.S. EPA, Section 301 (C) of CERCLA.
National Oceanic and Atmospheric Administration; 15CFR990 and Section 1006 (e)(1) of OPA.

Chapter 13
Environmental Risk Assessment Issues

Environmental risk management requires a thorough, comprehensive, and precise evaluation of environmental risk within the organization's activities as it enables the company to assess the risk inherent in accepting current practices. These evaluations are essential in determining the most appropriate environmental management options and the viability of standards and options.

This may also include assessing the net health impacts resulting from activities by subtracting the short-term environmental risks of the activity from the fatalities averted due to the reduction in long-term risks as a result of implementing an environmental control activity. In this manner, the risk analyses can evaluate the net health impacts at various cleanup levels and under various scenarios.

General Discussion

Baseline risk assessments can be based on individual exposures versus broader population-based risk analysis. The overall goal of risk assessments is twofold:

- Determine if a risk exists to the environment.
- Determine the level of risk to the environment, regardless of whether an individual exposure or population exposure technique is used. There are common elements to both techniques.

There are two types of risk assessments to consider: a health assessment and an ecological assessment. As to the type and level of detail required for a risk assessment, it is dependent on specific operating conditions and objectives.

Health Assessments

The objective of a health assessment is to provide:

- A basis for determining levels of chemicals that can be tolerated and still be protective of public health;

- An analysis of those baseline risks that generate a determination of the need for action;
- A basis for comparing potential health impacts of various alternatives; and
- A consistent process for evaluating and documenting public health threats at facilities.

The health assessment components include:

- Hazardous identification
- Dose-response assessment
- Exposure assessments
- Risk characterization

The first component of a health assessment, hazard identification, is the process of identifying which detected contaminants have inherent toxic effects and are likely to be of concern. It is based on a review of facility-specific monitoring and modeling information. The steps in making a hazard identification are as follows:

1. Determine the extent of contamination.
2. Calculate statistical means.
3. Evaluate non-detection and trace volume data.
4. Determine the background and naturally occurring values.
5. Select the contaminants of concern based upon concentration, toxicity, frequency of detection, sample location, and the preparation of the compound (chemical/physical).

The second component of a health assessment, a dose-response assessment, relates chemical exposure (dose) to expected health effects (response). The data generated by the dose-response assessment is evaluated relative to carcinogenic effects versus non-carcinogenic effects.

The third component of a health assessment, an exposure assessment, provides scientific information to evaluate the potential for public exposure to harmful dose levels. There are a few elements of an exposure assessment:

- Identification of exposure pathways
- Estimation of exposure-point concentrations for each selected pathway
- Estimation of exposure dose for each selected pathway
- Development of exposure scenarios

The exposure pathway is the key element in the exposure assessment and consists of four elements:

- A source and mechanism of chemical release into the environment;
- An environmental transport medium (a mechanism for the released contaminant to transfer from one medium to another);

- A point of potential contact with humans and biota; and
- A viable exposure route (air, groundwater, surface water, soil, food chain).

If all four elements are present, an exposure pathway is considered "complete;" if not, then the potential risk is diminished significantly.

Exposure-point concentration is defined as the amount of chemical in an environmental medium to which a person may be exposed. It can be expressed in either mass per unit volume (mg/l or mg/m^3) or unit weight (mg/kg). Exposure-point concentrations should be developed for each viable exposure pathway based on site sampling data or on modeling results.

The fourth component of a health assessment is risk characteristics. Chemical toxicity values, in conjunction with dose estimates for each of the various exposure pathways and population subgroups, can then be used to quantitatively estimate the carcinogenic health risks as well as the non-carcinogenic health risks.

Risk assessment draws heavily on the science of toxicology. *Toxicology* is the study of how toxic substances affect organisms. Central to these studies is the concept of dose and how it is expressed. All chemical substances can produce harmful effects if the difference between a toxic effect and no effect is the dose (and route of entry or exposure time). Typical routes of entry are inhalation, ingestion, absorption, and injection. The dose can be recorded in units of mg/kg of body weight for the oral dose; cm^2 for the skin dose; or parts per million (ppm), mg/m^3, and mg/l for the inhalation dose.

Toxicity is measured in terms of the dose-response relationship. A toxicity test exhibits a dose-response relationship when there is a consistent mathematical and biologically plausible relationship between a proportion of individuals responding and a given dose for a given exposure period:

(1) Acute — short-term, usually more intense
(2) Chronic — long-term, usually less intense
(3) Exposure — dose of a chemical
(4) Effects — body response to a chemical

} Can have an acute or chronic exposure with acute or chronic effects

Important dose-response terms are:

- NOAEL (No Observed Adverse Effect Level): The concentration or dose at which there is no adverse response in the population.
- LOAEL (Lowest Observed Adverse Effect Level): The lowest concentration that causes an adverse response in the population.
- TD50 (Toxic Dose 50): The concentration (or dose) that produces the adverse effect in 50% of the population (LD50 for Lethal Dose, if killing the population).

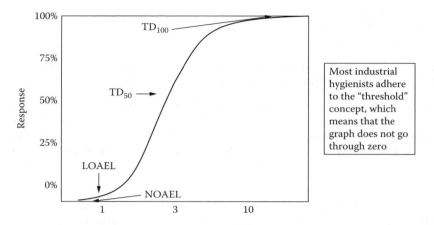

Exhibit 65. Dose response curve (dose, arbitrary units, logarithmic scale). Routes of entry: inhalation, ingestion, absorption, injection.

- TD100 (Toxic Dose 100): The concentration (or dose) that produces the adverse effect in 100% of the population (LD100 for Lethal Dose, if killing the population) as expressed in Exhibit 65, an example of a dose-response level.

Ecological Risk Assessment

The objectives of an ecological risk assessment are to assess the probability of adverse biological and ecological effects related to contamination from the operating practices of the company facilities. The assessment addresses the risk relative to past, present, and future site contamination impacts. The ecological risk assessment should play a key role in the development of cleanup criteria and the assessment of risk relative to remedial act alternatives.

There are five elements to an ecological risk assessment:

- Site characterization and identification of potential receptors;
- Selection of chemicals, species, and endpoints;
- Exposure assessments;
- Toxicity assessments; and
- Risk characterization.

The first ecological risk assessment element, site characterization and identification of potential receptors, involves establishing contamination, habitat, and species profiles. Relative to contamination, the researcher must establish the extent of contamination and the potential courses of contamination, and target the key contaminants of potential concern. In defining the habitat element to a given site, researchers must characterize species present, particularly endangered species and economically important species. It is also critical to establish what the potential indicator species are that reflect the overall ecological state of the site and provide a sound basis for the future site-monitoring program.

The second ecological risk assessment element is the selection of chemicals, species, and endpoints. The basis for selection of chemical contaminant criteria is the persistence of the contaminant, its high bioaccumulation potential, the given chemical's toxicity, and the potential for elevated levels at a given site above naturally occurring background levels.

The basis for the selection of species and endpoints criteria is:

- Their respective importance to the ecological system;
- Their sensitivity;
- The availability of practical methods for measurement and prediction; and
- The regulatory and trustee endpoint consideration.

Potential general endpoints include organisms, populations, and community. For organisms, endpoints may be mortality rates, changes in growth, changes in behavior, changes in structural development, reproductivity, impairment, mutagenicity, biochemical changes, and pathological abnormalities. From a population perspective, potential endpoints include species abundance, reproductive potential, and distribution. From a community perspective, potential endpoints for consideration include species composition, biomass, interspecies relationships, and extinction.

Exposure assessment is the third element of an ecological risk assessment. The objective of an exposure assessment is to identify:

- Which biological resources are exposed to chemical contaminants;
- Identify significant pathways and routes of exposure; and
- The magnitude, duration, and frequency of exposure.

The components of the exposure assessment include:

- Source characterization;
- Transport and fate analysis (i.e., migration mechanisms, spatial distribution, and temporal trends of contaminants);
- Exposure scenarios;
- Uncertainty analysis; and
- An integrated exposure assessment.

The latter entails establishing the population characteristics, the potential avoidance behavior, and the exposure-point concentrations of contaminants, as well as the duration and frequency of exposure.

Toxicity assessment is the fourth element of an ecological risk assessment. The toxicity assessment objectives are:

- Identify the potential toxic effects of the contaminants of concern.
- Identify the physical, chemical, and metabolic properties of each of the chemicals of concern.
- Determine the relationship between the amount of exposure to each chemical of concern and the resulting biological effect.

The elements of the toxicity assessment are hazard identification, the establishment of quantitative dose limits, and uncertainty analyses.

Hazard identification is based on the results of laboratory toxicity tests, field studies, and the quality review for the target endpoint in indicator species. Quantitative dose limits are the response data and toxicological indices for given species and compounds. These are based on laboratory toxicity tests for individual chemicals and complex mixtures as well as site-specific data.

The fifth and last element of an ecological risk assessment is risk characterization. Its objective is to determine the probability that adverse effects to the receptors of concern will result from the estimated exposure and to determine the degree of confidence in the risk estimate. The characterization of risk is based on:

- The estimated risks for single chemicals, single species, and specific endpoints;
- The multiple chemical risk predictions;
- The distribution of estimated risks;
- The risks to communities and ecosystems; and
- Uncertainty analysis.

Population Risk Analysis

As described previously, much of the early risk assessment modeling focuses on assessing risk to a *reasonable maximum exposed* (RME) individual. However, consideration is increasingly being given to considering the use of population risk as a risk descriptor in describing and communicating risks. Specifically, "The Policy for Risk Characterization" issued by Administrator Carol Browner in March 1995 states that:

> Agency risk assessments will be expected to address or provide descriptors of (1) individual risks that include central tendency and high end portions of the risk distribution, (2) population risk, and (3) important subgroups of the population, such as highly exposed or highly susceptible groups.

In addition, President Clinton's Executive Order 2866 on Regulatory Planning and Review included provisions designed to promote the use of risk analysis in making regulatory decisions in which benefits justify costs.

An aggressive use of population risk to define environmental program activities may result in the selection of more cost-effective (and implementable) operational solutions. Furthermore, a population risk approach may generate more statistically sound data.

The standard baseline risk assessments discussed in Chapter 9 summarize the relative contribution that each substance makes toward the total RME cancer risk and hazard index (non-cancer health risk) for each exposure medium. When calculating the RME for individual risks, the EPA uses the 95% upper confidence limit (95-UCL) on the arithmetic mean concentration as the assumed source of exposure for each substance. The concentration is used to provide an upper bound to individual exposures and risk. However, when estimating population impacts it is more appropriate to use a measure of central tendency like an arithmetic average or geometric mean concentration because most individuals in the population are not experiencing upper limit exposures over their lifetime. Whereas the total contaminant intake for a population is the sum of all individual intakes, some of the individual intakes will be somewhere in between depending on the frequency distribution of contaminant concentrations. Almost all frequency distributions of contaminant concentrations are log-normal, and in this type of distribution the geometric mean best represents the central tendency and is the most frequent value for each environmental medium. The difference between the geometric mean and 95-UCL can be quite significant.

The technical issues and methodologies used can be performed for (human) health and ecological assessments using relatively simple techniques and readily available information. In summary, population risks can be used to determine the most effective standard or operational approach based on health benefits and other factors, such as social, economic, and technical feasibility.

Whereas the individual cancer risk for residential exposure is almost always greater than the individual worker risk (because of factors such as intake rates, time on the property, and the large number of exposures), the reverse is typically found under a population risk analysis. This is due to the fact that the industrial population, assumed to be working on the property, is typically much greater than the number of residents living on or near the site. The greater the number of individual workers compared to the number of affected residents more than complements for the greater individual residential exposure. Thus, the results of the population risk assessment can provide additional insight for the operational control remedy selection process that cannot be realized by analyzing individual risk alone.

References

"Choices in Risk Assessment: The Role of Science Policy in the Environmental Risk Management Process." Prepared for Sandia National Laboratories, 1994.

DOE, June 1989. "A Manual for Implementing Residual Radioactive Material Guidelines" (DOE/CH8901).

Dragun, James. *The Soil Chemistry of Hazardous Materials* (Hazardous Materials Control Research Institute, 1988).

EPA, Office of Air and Radiation. "Technical Support Document for the Development of Radionuclide Cleanup Levels for Soil (Review Draft)." September 1994.

EPA, Office of Emergency and Remedial Response. "Population Risk Analysis for Superfund Sites Using Simple Techniques and Readily Available Information." Prepared by SC&A, March 1998.

EPA, Office of Research and Development. "Update to Exposure Factors Handbook (Draft Report)." 1996b (August 1996).

EPA, Office of Solid Waste and Emergency Response. "Soil Screening Guidance: Technical Background Document" (EPA/540/R-95/128, 1996a (May 1996)).

Hoffman, F.O. and C.F. Bates, III. "A Statistical Analysis of Selected Parameters for Predicting Food Chain Transport and Internal Dose of Radionuclides" (NUREG/CR-1004, 1979).

Howard, Philip H., Robert S. Boethling, William F. Jarvis, William M. Meylan, and Edward M. Michalenko. *Handbook of Environmental Degradation Rates* (Lewis Publishers, Inc., 1991).

Schaffer, S.A. "Environmental Transfer and Loss Parameters for Four Selected Priority Pollutants." Proceedings of the National Conference on Hazardous Waste and Environmental Emergencies, May 1985, Cincinnati, Ohio.

Chapter 14
Emergency Response Analysis

Emergency response analysis is a decision-making tool to identify potential emergencies facing a community with respect to accidental explosions or releases of compounds from company facilities. By identifying and evaluating all potential emergencies—including the highly unlikely worst-case scenarios—decision makers are able to prepare for emergencies and significantly reduce the risks. A population risk-based approach should be developed to assess a variety of land use, exposed population, and exposure pathways, as applicable to a given site. In the emergency response analysis context, safety is the reciprocal of risk, or the probability that substances will not produce harm under the same conditions. Thus, when ultimately determining the risk or safety of the present approach the critical factor will not necessarily be the intrinsic hazard potential of the substances per se but the safety precautions that have been taken to decrease the potential harm from that substance.

It is assumed that the greatest acute hazard potential to the nearest community will be via air-borne transmission in the form of a vapor, gas, or dust cloud, or via an explosion of a volatile container. Using these assumptions, worst case scenarios for representative compounds should be conducted to indicate the size of the hazard zones that would have to be considered in emergency planning for the present approach.

The ARCHIE Model series was developed by the EPA, the Department of Transportation (DOT), and the Federal Emergency Management Administration (FEMA) for use in hazard analysis. EPA's Automated Resource for Chemical Hazard Incident Evaluation (ARCHIE) model is the recommended model in several states as part of the emergency response analysis requirement in permit requests. It should be stressed that the assumptions are considered to be worst case, i.e., the situation is assumed to be the worst that can occur based on the information. As such, severe combinations of factors (i.e., temperature, wind, and others) have been fed into the model than would normally be encountered at a given site that results in exaggeration of the size of hazard zones. Some examples of model components are briefly discussed below.

The condensed-Phase Explosion Model assesses the physically destructive impact of a potential explosion. The purpose of the Vapor Dispersion Model is to provide an estimate of the dimensions of the initial downwind zone that may require protective action in the event of a hazardous material discharge due to the accidental discharge, emission, or release of a toxic gas or vapor to the atmosphere. The worst case for vapor dispersion would be the substance with the highest vapor pressure and lowest immediately dangerous to life and health (IDLH) dose. This is because the model assumes that all the contents of the drum will be released into the atmosphere in one minute, and thus vapor pressure does not play a significant role.

The Release Pressurized Gas Pipeline Model provides an estimate of hazard zones if a pipeline with gas under pressure is ruptured during cleanup activities at a site.

Contingency Plans

OSHA Regulation 29CFR 1910.120 requires written emergency response plans. Contingency plans and emergency response plans are step-by-step guides for handling emergencies and include a description and identification of the identification and location of hazards, roles and responsibilities, available equipment, evacuation procedures, and decontamination procedures. It also calls for establishing related issues such as safe distances, security of a site, PPE, methods of alerting on-site personnel, and notifications and reporting requirements off-site.

EPA Regulation 40 CFR 355.10 and 372 also requires written emergency response plans; from the EPA's standpoint, contingency plans and emergency response plans are step-by-step guides for handling emergencies. They include a description and identification of response capabilities. These include:

- Levels of emergencies—standardizes what type of response is needed and who will be responding
- Notifications—requirements for documentation and record-keeping of information provided

Planning mitigates the potential for a full emergency to develop. By conducting risk assessments and analyzing the potential for hazardous activities to impact employee safety, you can identify the unsafe acts that lead to accidents and emergencies. Planning for an emergency reduces the likelihood of it occurring. Pre-incident planning includes knowing the hazards present, the resources available, and what can reasonably be accomplished during an incident.

By knowing what hazardous materials are present, the emergency officer can estimate the likely harm if not controlled, establish response objectives, and weigh offensive or defensive options. Another reason for pre-planning is to manage the potential outcome of an event. By knowing the resources available and by understanding what can be accomplished, the emergency manager has a better chance to manage the outcome of any situation (magnitude, occurrence, timing, effect, or location).

Up-the-Stack Emergencies

One area that requires particular vigilance is the reporting of accidental and under-estimated air emission release from industrial facilities. "Up-the-stack" industrial companies may underestimate emissions for three reasons:

- Companies are only required to report releases above certain amounts under both federal and state law. For example, companies must report the release of carbon monoxide only if more than 5000 pounds is emitted in a 24-hour period. If frequent but smaller releases occur, they may not be captured because there is no reporting required.
- Designs typically will call for gases to be routed to a temporary control feature (e.g., a flare). Company emissions reports usually assume the control feature is operating at maximum efficiency, i.e., destroying a significant percentage of emissions. But there are times when the secondary capture is incomplete and therefore not operating at designed efficiency. As a result, the release may be more than the reported estimates of pollution into the atmosphere.
- Some reports do not state the amount of pollution released at all but simply note that the reportable quantity has been exceeded.

The unpermitted release of smog-forming volatile organic compounds (VOC) is of special concern for up-the-stack industrial organizations. In particular, facilities in communities that are in non-attainment areas that do not meet health-based standards for ozone must show strong care with regard to unpermitted releases. High levels of ozone in the atmosphere can result in several known health effects, including irritation of the respiratory system, reduction of lung capacity, aggravation of asthma, and inflammation and damage of the lung lining. Scientists also suspect that ozone may aggravate chronic lung diseases, such as emphysema and bronchitis.

In addition to elevating the risk of cancer, lung disease, and other ailments associated with long-term exposure to the pollutants released from up-the-stack industrial facilities, the release of high volumes of pollution in a short period of time can trigger acute health effects. This may include asthma, nausea, depression of the nervous system, heightened cardiac sensitivity, and heart attack. As an example, acute exposure to high concentrations of benzene (2000 ppm) can depress the nervous system and cause death, whereas exposure to lower concentrations (1000 ppm) may cause nausea, headaches, heart arrhythmia, anemia, and blood cancers such as leukemia.

A good starting point for strong community relations is to develop a system for notifying residential neighborhoods when accidents occur or what is being done to stop them. Information contained in an individual release report should include the pollutants released and the amount and should be readily available by making a personal visit to the local facility's office.

All accidental releases, as well as many that result from maintenance or shutdown activity of hazardous chemicals above a specified amount, must be reported to the federal government's National Response Center within 24 hours. However, it is not always clear from company reports whether startup or shutdown releases meet these exemptions, or whether the shutdown was a planned activity or part of the emergency response to an accident. Whereas federal guidance requires that unanticipated emissions from accidents should always be reported, releases from startups and shutdowns do not have to be reported if the releases are subject to federally enforceable limits or pollution controls and are part of a plan approved by the permit authority.

While the Clean Air Act regulations excuse the release of thousands of pounds of even a cancer-causing pollutant like benzene if it is beyond the reasonable control of the plant operator and meets other conditions, it does not make sense to rely on this loophole in one's environmental management practices. Releases that result from a sudden breakdown of process or control equipment—as well as from regularly scheduled startup, shutdown, or maintenance activities needed to cope with predictable wear and tear—seriously impact communities and community relations. Therefore, they should be rigorously controlled. And the EPA does consider these emissions illegal as they generally exceed established permit limits. However, at the same time the EPA and states generally do not seek penalties for these types of releases as long as they are beyond the control of the operator. But any facility operator who relies on this to ensure environmental management plan is unwise and leaving the company and the board at great risk.

But this "beyond the control defense" does not provide companies with a free pass. The EPA and state agencies have the authority to order plant operators to investigate and fix the underlying cause of the accident by forcing the facility to install better pollution control equipment. Moreover, to avoid penalties a company often has the burden of proof to show that the accident could not have been prevented and that all steps were taken to minimize emissions.

In cases involving a startup or shutdown, the facility must prove that "the periods of excess emission that occurred during startup and shutdown were short and infrequent and could not have been prevented through careful planning and design." In addition, the excess emissions cannot be "part of a recurring pattern indicative of inadequate design, operation, or maintenance." Most importantly, the facility must show that "all possible steps were taken to minimize the impact of the excess emissions on ambient air quality."

Regarding malfunctions, the defendant carries the burden to prove that:

> The excess emissions were caused by a sudden, unavoidable breakdown of technology, beyond the control of the owner or operator...[t]he excess emissions (a) did not stem from any activity or event that could have been foreseen and avoided, or planned for, and (b) could not have been avoided by better operation and maintenance practices...[and] [t]he excess emissions were not part of a recurring pattern indicative of inadequate design, operation or maintenance.

While the EPA warns that accidents that should have been anticipated and prevented will not be excused under the Clean Air Act, there is a sense of discomfort that the exceptions may lead to a "Russian roulette" behavior on the part of negligent operators and little defense for the public should significant releases or accidents occur. The fact is that these problems can be mitigated if not fixed.

Improved technology and work practices can eliminate pollution from accidents, startups, and shutdowns. Because accidents are inevitable, facilities should incorporate a variety of practices and technologies to minimize the effects of an accidental release. An important first step is better management of the flow of raw materials to avoid overwhelming production units and triggering shutdowns or emergency upsets. Also, facilities could recycle VOCs back into the manufacturing process through a closed-loop system; add temporary storage capacity for all waste gases normally flared; and build redundancies and backup systems, including triple backup or redundant systems for electronic controls or major compressor units and other sensitive equipment that can fail due to false electronic glitches.

There should be requirements to diagnose the root causes of malfunction and emergency releases. Companies can improve backup pollution controls used when equipment goes down. Board of directors beware: the public will not be tolerant of accidents that result from not incorporating these practices into their manufacturing processes when they can stop the continued unpermitted release of air pollutants from accidents, as well as startup, shutdown, and maintenance activities. In short, accidents will happen even at the best-run plants, but they should not become a way of life.

In summary, four actions should be taken to protect corporate and community interests in the event of an accidental release:

- Better reporting of emissions;
- Stronger monitoring of air quality;
- Stronger accident prevention planning; and
- Lessen reliance on the loophole for unpermitted releases.

General Emergency Management Concepts

Potential weaknesses often found in incident management include:

- Lack of personnel accountability
- Poor communication
- Lack of a planning process
- Poor methods for integrating assets

The Incident Command System (ICS) should be based on the best practices of standard emergency management systems around the country.

The ICS should be interdisciplinary and organizationally flexible. It should be flexible enough to meet the needs of any kind of size incident. It should be flexible enough to be used for both routine and planned and complex emergency incidents; the latter may include major natural disasters and acts of terrorism. It should the basis for allowing personnel from a variety of organizational settings to quickly meld into a common, cost-effective organizational structure.

The ICS is a standardized management approach designed to allow users to adopt an integrated organizational structure flexible enough to meet the demands of small or large emergency situations. Identified areas of management weaknesses resulted in the development of the ICS best practice approach.

Recognize that there is often no correlation between the needs of the ICS organization and the company's administration structure. Regardless of the size or type of emergency, every incident or event requires certain management functions to be performed. The problem must be identified and assessed, a plan quickly developed to deal with it, and necessary resources must be identified and provided.

There are five basic management functions or sections to be established within the ICS organizational structure:

- Incident Command—sets objectives, strategy, priorities, and has overall management responsibilities over the incident.
- Operations—conducts the tactical operations, develops tactical objectives and organization, and directs tactical resources.

- Planning—prepares and documents the Incident Action Plan (IAP) to accomplish objectives and collect and evaluate information, as well as maintaining resource status, documentation, and records.
- Logistics—provides support resources and services needed to meet operational objectives.
- Financial/Administrative—tracks costs relative to the incident and provides accounting, procurement, time records, and cost analysis support.

The Incident Commander (IC) is the only position that is always filled. For a small incident, the IC can cover all five ICS roles. Each primary ICS section can be expanded or contracted to meet the needs of the incident. However, basic ICS operational guidelines are that the person at the top is responsible until authority is formally delegated to another person. It has established that an optimal command span of personnel control during emergency response is on the order of three to seven, with five being optimal. Maintaining an effective span of control is particularly important on those incidents where safety and accountability are critical.

It is also critical to develop and use specific ICS organization position titles and responsibilities that are well recognized by all parties. Rank grade and seniority should not be factors used to select an IC and ICS section positions. These positions should be filled by the most highly qualified individual trained to lead the incident response. Any transfer of command requires a full briefing for the incoming IC and notification to all personnel of change in command.

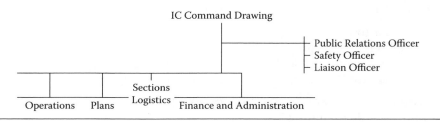

The Operations section chief will develop and manage the Operations section to accomplish the incident objectives set up by the incident commander. The Operations section chief is normally the person with the greatest technical and factual expertise in dealing with the problem at hand.

Supervisors within the Operations section may be established based on:

- Division (geographic)
- Group (functional)
- Branches (either geographic or functional)

The above may be established based on the span of control requirements, multidisciplinary issues, multijurisdictional, issues and the size of the incidents.

Managing operations may involve developing the following:

- Task forces—combinations of mixed resources with common communication and operating under the direct supervision of a task force leader.
- Strike teams—a set number of resources of the same type that have common communication and are operating under the direct supervision of a strike team leader.
- Single resources—individuals or equipment that complement that incident response organization in a specialty fashion.

The role of the Planning section may include the following:

- Collect, evaluate, and display incident intelligence information
- Prepare and document the IAP
- Conduct long-range contingency planning
- Develop the plan for demobilization
- Maintain incident documentation
- Track resource assignments

The Incident Action Plan (IAP) should be prepared for every incident. The purpose of the plan is to provide all incident personnel with direction for the actions to be implemented during the operational period identified in the plan. Typically, IAPs should be prepared on a daily basis and address:

- What to do
- Who is responsible
- How are communications handled with other organizations
- What are the procedures for handling outside inquiries

The Logistics section chief is responsible for all the servicing and support needs including:

- Obtaining, maintaining, and accounting for essential personal equipment and supplies
- Providing identified communication resources
- Providing food services, if necessary
- Setting up and maintaining incident facilities
- Providing support transportation
- Providing medical services as necessary

Typically, the logistics chief will organize his resources into two units:

- Service Branch (communication, medical, food)
- Support (supplies, facility management, general)

The Financial/Administration section is set up for any incident that requires incident specific financial management. These may include responsibilities for:

- Timekeeping
- Contract negotiation and monitoring
- Cost analysis
- Compensation for injury or damage to property

The ICS addresses the establishment of incident command structure, the need for authorities, and delegation of authorities. It includes operation plans, memorandums of understanding (MOU) with local authorities, and the development of the IAP. It also addresses the needs for transition briefings if transfers in command must occur during the incident, as well as briefings to senior corporate management. The information needed to conduct these briefings is similar and should include the extent of damage, probable response, needs, and resources on the scene and their locations. The briefings may also include safety concerns, political issues, and other concerns. Reasons for the transfer of command briefings and senior corporate officer briefings are to:

- Identify jurisdictional (legal) issues and boundaries
- Understand the complexities of changing environments
- Understand senior management-directed change
- Understand long-incident duration fresh staffing requirements
- Update on staff morale and energy

The tactical incident action planning process is to assist senior corporate management and the incident management team in the systematic and orderly development of a plan of action in a minimum amount of time. It is recommended that the planning process be broken into daily or 12-hour increments as the urgency of the situation dictates. The goals of the process are to:

- Understand overall goals, policy, and direction
- Assess incident situation on a timely basis
- Quickly establish and refine incident objectives as realities are understood
- Quickly select and obtain approvals for appropriate strategies to achieve objectives
- Perform tactical direction (applying tactics appropriate to strategy, assigning the right resources, and monitoring performance)
- Provide necessary follow-up (change strategy, tactics, add/subtract resources, and so on)

It is critical that there be a well-defined, unified command regardless of whether organizationally it is a unilateral, shared, or joint responsibility incident. There should be one supervisor to whom incident staff reports.

However, a realistic chain of command should be set up to recognize the span of control limitations.

The ICS approach is based on a management-by-objectives philosophy. It recognizes that every incident has different requirements. ICS emphasizes that the organizational structure should reflect only what is required to meet and support the planned incident objectives.

Effective ICS communication includes:

- Establishing procedures and processes for transferring information internally and externally
- Providing hardware systems to transfer information rapidly
- Identifying and developing a plan for using all available communication resources

Preparedness for incident management can be attained by:

- Planning and training exercises
- Establishing and verifying personnel qualifications and certification standards
- Engaging in pre-planning equipment identification, sourcing, acquisition, and certifying
- Developing and publishing management policies and processes
- Identifying and establishing mutual aid agreements and emergency management assistance compacts requirements
- Managing the incident at the lowest geographical/organizational level possible

References

Environmental Integrity Project. "Accidents Will Happen—Pollution from Plant Malfunctions, Startups, and Shutdowns in Port Arthur, Texas." (Washington, D.C., 2002).

EPA, Office of Acid Deposition, Environmental Monitoring, and Quality Assurance. "The Total Exposure Assessment Methodology (TEAM) Study: Summary and Analysis," Volume 1 (EPA/600/6-87/002a).

FEMA. NEMIS Management Training, IS-00100.FW, IS-00200.FW, IS-00700.FW. (Emergency Management Institute, 2006).

Chapter 15
Corporate Health and Safety System

Companies face increasing public expectations to ensure the adequacy of the health and safety of their employees. The effectiveness of these arrangements is an important contributor to the overall safety performance of the company, and is also increasingly seen as an indicator of potential community impacts. These efforts have become more formalized and structured through the introduction of legislation, the application of risk assessment and audit procedures to assess a widening range of hazards, and the concomitant development of standards and guidance.

Similar to environmental management systems, occupational health and safety systems are geared to developing uniform standards versus a system tailored to an individual firm. Regardless, a comprehensive occupational and safety system includes the following elements:

- The formulation of an occupational health and safety policy;
- The identification of risks and legal requirements;
- Establishment of objectives, targets, and programs that ensure continual improvement;
- Evidence of management activities to control occupational health and safety risks;
- Monitoring of the system's performance; and
- A policy of continual reviews, evaluation, and improvement of the system.

The development of standardized occupational health and safety systems is a new initiative component of environmental standards. In 1996, the British Standards Institute launched the world's first standard, the "BS 8800: Guide to Occupational Health and Safety Management Systems." It was later revised to incorporate ISO 14001 and promulgated as "OSHAS 18001: Occupational Health and Safety Assessment Series." OSHAS 18001 is the preferred standard in most industries.

Implementation of occupational health and safety systems is a relatively recent development as a standardized tool, and only a limited number of companies in most countries have so far implemented formal occupational health and safety systems. Most companies rely on their own tailored solutions that in part address these issues but are not in a manner that invites systematic comparison. Evidence of a health and safety system is focused on three criteria:

- A senior company official is designated as responsible for occupational health and safety;
- Details of health and safety training programs are readily available; and
- Detailed quantitative data is generated to illustrate performance in a constant and comparable fashion.

The OECID/EIRIS study found that just under half of the companies in the sample display at least some evidence of having an occupational health and safety system in place. The lowest incidence of health and safety systems is found among Asian firms and the highest is found in Europe.

Establishing Hazard and Safety Control Measures

There are numerous chemical, physical, environmental, and even at times radiological hazards that can potentially present at operating facilities. If not properly controlled, these hazards can cause harm to project personnel, visitors, and the public. The anticipated hazards and the recommended control measures need to be identified and addressed in detail in site health and safety plans per the Occupational Safety and Health Administration (OSHA) regulations for Hazardous Work Operations and Emergency Response (HAZWOPPER, 29CFR1920.120).

Historical process information can be used to indicate the presence of contaminants and other hazards of concern. Typically, there is potential for exposure to operational personnel through various routes (dermal contact, inhalation, ingestion, injection). Controls must be specified in health and safety plans to reduce the risk of these potential exposures.

To minimize the risk of potential exposure of employees to hazardous chemicals, it is important to understand how personnel can be affected through exposure. There are three main sources of exposure: inhalation, ingestion, or absorption. Once the route has been established, it is important to distinguish between the type of damage (localized or systemic). *Systemic damage* addresses the broader effects of the chemical, and may include target organs, whereas *localized damage* appears at the point of contact. The body's biological response to the amount of exposure is called the *dose-response relationship*. Dose is expressed in parts per million (ppm), parts per billion (ppb), and parts per trillion (ppt). Skin contact may be expressed in mg/m^3. Exposure may be acute or chronic. Safe limits are established based on the toxicity of the material, the duration of the exposure, and the dose-response relationship identified.

A brief definition of important inhalation exposure terms is provided below:

- *Threshold Limit Value—Time-Weighted Average (TLV-TWA).* Airborne concentrations of substances are generally expressed as an eight-hour TWA and represent conditions under which it is believed that nearly all workers may be repeatedly exposed day after day for a 40-hour work week without adverse health effects. TLVs are guidelines for occupational exposures established by the American Conference of Governmental Industrial Hygienists (ACGIH, 1998), and should be used only on controlled sites where contaminants and concentrations are well known.
- *Threshold Limit Value—Short-Term Exposure Limit (TLV-STEL).* The STEL is the concentration to which it is believed that workers can be exposed continuously for a short period of time without suffering from irritation, chronic or irreversible tissue damage, or narcosis of sufficient degree to increase the likelihood of accidental injury, to impair self-rescue, or to materially reduce work efficiency, provided that the daily TLV-TWA is not exceeded. An STEL is defined as a 15-minute TWA exposure that should not be exceeded at any time during the work day, even if the eight-hour TWA is within the TLV-TWA. Exposures above the TLV-TWA up to the STEL should not be longer than 15 minutes and should not occur more than four times per day. There should be at least 60 minutes between successive exposures in this range.
- *Recommended Exposure Limit.* The up-to-10-hours per work day TWA exposure limits are recommended by the National Institute of Occupational Safety and Health (NIOSH).
- *Immediately Dangerous to Life or Health (IDLH).* The IDLH is a concentration that poses an immediate threat to life or produces irreversible, immediate debilitating effects on health (American National Standards Institute). NIOSH defines IDLH as air concentrations that represent the maximum concentration from which, in the event of respirator failure, one could escape within 30 minutes without a respirator without experiencing any escape-impairing or irreversible health effects.
- *Permissible Exposure Limit (PEL).* The PEL is the eight-hour TWA, STEL, or ceiling concentration above which workers cannot be exposed. These enforceable standards are by OSHA.

Inorganic Chemicals

Various inorganic chemicals—specifically, metals—can be considered toxic, and some are identified as being carcinogenic. Detection analysis for each contaminant of concern should be presented in the Health and Safety Plan. For example, arsenic is a toxic, gray, brittle metal that may injure multiple organs. Acute injury usually involves the blood, brain, heart, kidneys, and gastrointestinal tract. The bone marrow, skin, and peripheral nervous system may develop chronic toxicity after acute or chronic exposure. Thus, an acute ingestion may cause both acute and chronic syndromes. The ACGIH has listed arsenic as an A1, Confirmed Human Carcinogen. (PEL: 0.010 mg/m^3, IDLH: 5 mg/m^3, TLV-TWA 0.010 mg/m^3) TLV Basis-Critical Effect(s): Cancer (lung, skin).

Organic Compounds

Organic compounds (hydrocarbons) may also be present as contaminants in the soil. Additional information about these chemicals should be found in the Material Safety Data Sheets (MSDS) kept on-site. A listing of the available MSDSs should be maintained at the health and safety field office, and a description of potential concerns addressed in the Health and Safety Plan. For example, hydrocarbons are a group of semi-volatile organics that are rather persistent in the environment. Some polynuclear aromatic hydrocarbons (PAH) are carcinogenic with inhalation as the primary exposure route. The greatest carcinogenic effect is at the point of contact (i.e., lungs, skin, and stomach). Skin disorders may also result due to high concentration exposures. Exposure limits have not been established for many specific PAHs in this large group of compounds.

Operational Chemicals/Hazard Communication Program

The use of operational chemicals is regulated by OSHA under the "Hazard Communication Standard" (29CFR1910.1200). Air monitoring must be performed as needed to assess exposures resulting from their use. MSDSs for operational chemicals must be kept on file at all company operational facilities and an inventory list of the anticipated operational chemicals (Hazardous Chemical Inventory List) for use must be maintained at the facility.

The tools provided under the regulations to identify and classify materials include MSDS, labels, shipping papers, classification criteria, and other identifying markers to express the hazards associated with the material.

Hazardous communications is the employer's responsibility. The employer must provide the employee with the information about a chemical that they may be working with or around as per the OSHA "The Employee-Right-to-Know" program. Whereas it is the employer's responsibility to provide the information and training on the specific chemicals in the work place, the employee is responsible to understand and keep current on the different chemicals they use. Below are some of the issues and responsibilities that must be understood at all times by all employees:

- Where the MSDSs are kept
- How hazardous is the chemical in general terms

- What to do in the advent of an emergency (personal contamination, first aid, spill response responsibilities, evacuation)
- What personal protective equipment is required

The Department of Transportation (DOT) classification information includes the nine classes of hazardous materials, placards, labels, shipping papers, and manifests:

- Class 1—Explosive
- Class 2—Flammable Gases
- Class 3—Flammable Liquids
- Class 4—Flammable Solids
- Class 5—Oxidizers
- Class 6—Poisons
- Class 7—Radioactive Materials
- Class 8—Corrosive
- Class 9—Miscellaneous

Documentation includes manifests, consist logs, inventories, shipping papers, placards, and labels.

The EPA classifies wastes as listed or characteristic. *Listed wastes* are specific wastes from a specific source, or specific wastes from a non-specific source. *Characteristic wastes* are defined as Ignitable, Corrosive, Toxic, and Reactive or Instability. OSHA requires every employer to provide training to employees on the hazards associated with chemicals used in the work place before handling chemicals. This includes understanding MSDSs and warning labels.

Other important terms and concepts of chemical hazards include fire/flammability and flammable or explosive limits. For fire/flammability to be a concern, three elements that must be present are fuel, heat, and oxygen. Flammable or explosive limits are measured in terms of a flammable range bounded by the lower explosive level (LEL) or lower flammable limit (LFL) and the upper explosive level (UEL) or upper flammable limit (UFL). Exhibit 66 provides an example.

Personal Protective Equipment

When engineering and administrative controls are not feasible or not adequate to protect personnel from the hazards associated with facility operations activities or energy clean-ups, *personnel practice equipment* (PPE) must be required.

Respiratory Protection. When deemed necessary, a respiratory protection program should be implemented that is compliant to the requirements of 10 CFR 20 Subpart H, "Respiratory Protection and Controls to Restrict Internal Exposure in Restricted Areas," and EM825-1-1 06.E.07, "Respiratory

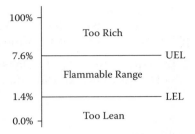

Exhibit 66. UEL/LEL example for gasoline.

Protection and Other Controls." Respiratory protection equipment must be NIOSH-approved and respirator use must conform to ANSI Z88.2 and OSHA 29 CFR 1926.103 requirements. These documents detail the selection, use, inspection, cleaning, maintenance, storage, and fit testing of respiratory protection equipment.

Levels of Protection. PPE is used as a last line of defense to control employee exposure to hazardous chemicals. PPE must be selected based on the hazards identified, must be appropriate for the degree of hazard, and employees must be trained on the selection, use, care of, and advantages and disadvantages of the PPE.

Eye Protection. In areas where there is the potential for flying objects to enter the eye—dust, mist, fumes, or vapors—eye protection is required.

- Safety glasses
- Safety goggles
- Face shields
- Visors

Hand Protection. Anywhere there is the potential for cuts, abrasions, punctures, chemical burns, thermal burns, or harmful temperatures, hand protection must be offered.

- Fit
- Types of gloves
- Barrier creams

Chemical Protective Clothing. Required when the employee has the potential exposure to airborne contaminants, splashing, spilling, or other activities where full body contact is possible, chemical protective clothing must be worn.

- Aprons/bibs
- Suits
- Levels of protection (Exhibit 67)—Level A, Level B, Level C, Level D

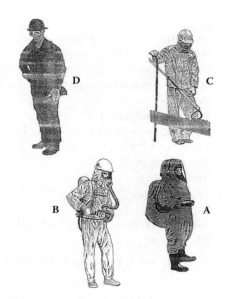

Exhibit 67. Levels of protection: Level A, Level B, Level C, and Level D.

Respiratory Protection. Employees with potential exposure to dust, fumes, mist, vapors, or sprays must be provided respiratory protection if engineering controls or administrative controls are not feasible.

- Dust masks
- Air purifying respirators
- Supplied air

Hearing Protection. Employees exposed to continuous noise at or above 85 dB for an eight-hour TWA must be provided with hearing protection and enrolled in a hearing conservation program.

- Earplugs
- Earmuffs
- Attenuators

Level A Protection. If used, Level A protective equipment shall consist of an enclosed self-supplied air respirator with personnel in a chemically compatible enclosed working suit (i.e., moon suit) and boots with an air-tight splash shield assembly (Exhibit 67.) Level A should always be used when the expected concentrations are at or near IDLH.

Level B Protection. If used, Level B protective equipment shall consist of (Exhibit 67):

- Supplied respirator
- Work clothing (light or insulated) as prescribed by weather

- Steel-toed boots
- Chemical resistant boot covers or outer boots, as selected by a certified industrial hygienist (CIH)
- Tyvek® coveralls with hoods or an equivalent protective garment with elastic wrists and ankles (or equivalent cloth/synthetic fiber), as determined by the safety officer
- Acid gear, splash suit, rain gear, and so on, as determined by a CIH
- Nitrile, latex, or vinyl gloves (inner) or cloth liners
- Outer gloves, as selected by a CIH
- Hearing protection (if necessary)
- Cooling vest (if necessary)
- Hard hat
- Splash shield (if necessary)
- Openings at ankles, wrists, and hoods shall be taped, as directed by the facility's safety officer

Level C Protection. If used, Level C protective equipment shall consist of (Exhibit 67):

- Full-face air purifying respirator (APR) with NIOSH-approved combination high-efficiency particulate air/organic vapor cartridges
- Work clothing, as prescribed by weather
- Steel-toed boots
- Chemical resistant boot covers or outer boots—polyvinyl chloride (PVC)/Latex/Neoprene
- Tyvek® coveralls with hoods and elastic wrists and ankles (or equivalent cloth/synthetic fiber), as determined by the safety officer
- Nitrile, latex, or vinyl gloves (inner) or cloth liners
- Nitrile gloves or PVC (outer) or leather palm gloves
- Hearing protection (if necessary)
- Cooling vest (if necessary)
- Hard hat
- Splash shield (if necessary)
- Openings at ankles, wrists, and hoods shall be taped, as directed by the facility's safety officer

Level D-Modified Protection. Level D-Modified PPE can consist of the minimum Level D plus any of the additional items listed below:

- Work clothing, as prescribed by weather
- Chemical resistant boot covers, totes or equivalent (PVC/Latex/Neoprene)
- Tyvek® coveralls with hoods and elastic wrists and ankles (or equivalent cloth/synthetic fiber), as determined by the facility's safety officer
- Nitrile or vinyl gloves (inner) or cloth liners
- Nitrile or PVC gloves (outer) or leather palm gloves
- Hearing protection (if necessary)

- Splash shield (if necessary)
- Cooling vest (if necessary)
- Openings at ankles, wrists, and hoods shall be taped, as directed by the facility's safety officer

Level D Protection. Level D protection is the minimum level of protection that will be used at an operational facility and is the typical operating level. At a minimum, Level D PPE shall consist of (Exhibit 67):

- Steel-toed work boots
- Safety glasses
- Hearing protection (if necessary)
- Hard hat
- Splash shield (if necessary)
- Leather work gloves (as necessary)

Monitoring and Medical Surveillance. Monitoring is done to verify the absence or presence of hazardous materials in the work environment. A medical surveillance is performed to verify the absence or presence of employee exposure to hazardous chemicals.

Monitoring. Monitoring can be done both for area contaminants and for employee exposure (personal monitoring).

- Area monitoring—looking at atmospheric conditions (explosive levels, oxygen levels, organic vapors, and so on)
- Personal monitoring—looking for potential exposure to employees
- Background monitoring
- Periodic monitoring
- Post-incident exposure monitoring

Measurement Instruments. There are two general approaches used to identify or quantify airborne contaminants:

- On-site use of direct-read instruments; and
- Lab analysis of samples taken.

The advantage of direct-read instruments is that they provide real-time data. Disadvantages of direct-read instruments include their limits in detecting/measuring of specific classes of chemicals. They are not typically designed to detect less than 1 ppm and are subject to interference problems. Direct-read instruments are:

- Combustible gas meter
- Oxygen meter
- Flame ionization detector
- Photo-ionization detector
- Colorimetric tubes
- Gas-specific instruments

- Radioactivity detectors
- Particulate detectors

Other various monitoring devices typically used include:

- Oxygen meters
- Organic vapor monitors
- Combustible gas indicator (CGI)
- Colorimetric tubes
- Geiger-Mueller pancake probes
- NaI scintillation meter

Site Control of Work Zones

Site control requires the designation of work zones as required by 10 CFR 20 Subpart J, specifically 1901—Caution Signs; 1902—Posting Requirements; and 1904—Labeling of Containers. These requirements are mirrored in EM 835-1-1 06.E.08, "Signs, Labels, and Posting Requirements."

If chemical contamination exists, work zones will be divided, as suggested in "Occupational Safety and Health Guidance Manual for Hazardous Waste Site Activities," NIOSH/OSHA/U.S. Coast Guard/USEPA, November 1985 into three zones: Exclusion Zone, Contamination Reduction Zone, and Support Zone.

Exclusion Zone (EZ). All employees are required to follow established procedures, such as wearing the proper PPE, when working in these designated areas. An entry log should be kept daily that records the time of entry and exit from the area for each person.

Decontamination of equipment and personnel may be necessary in controlled areas to reduce worker risks. Decontamination will generally occur at the edge of an area. Everything that enters a restricted area at the site must either be decontaminated or properly discarded upon exit. Everything that leaves a restricted area must be frisked to determine if contamination is present, and if it is, it should either be decontaminated or properly discarded.

References

ACGIH. "Threshold Limit Values and Biological Exposure Indices." 1999.
EPA. Safety Operating Guidelines (July 1988).
National Safety Council. "Fundamentals of Industrial Hygiene," 1996.
NIOSH. "The Effects of Workplace Hazards on Male Reproductive Health." (DHHS Publication No. 96-132).
Nuclear Regulatory Commission. Regulatory Guide 1.86, 1974.
OECI Secretariat and EIRIS. "An Overview of Corporate Environmental Management Practices, Joint Study by the OECD Secretariat and EIRIS."

Title 10 CFR, Part 19, "Notices, Instructions and Reports to Workers: Inspection and Investigations."

Title 29 CFR, Part 1910, "Safety and Health Regulations for General Industry."

Title 29 CFR, Part 1926, "Safety and Health Guidance Manual for Hazardous Waste Activities." (NIOSH Publication No. 85-115, October 1985).

USACE. Safety & Health Requirements Manual (EM 385-1-1, 3 September 1996).

U.S. Department of Health and Human Services, Public Health Service, Centers for Disease Control and Prevention, National Institute for Occupational Safety and Health (NIOSH, et al.), "NIOSH Pocket Guide to Chemical Hazards," (NIOSH Publication No. 97-140, June 1997).

Chapter 16
Environmental Risk Management at Banking Institutions

Potential environmental liability is a growing influence in the banking industry. In response, banking institutions are increasingly adopting environmental risk management programs. One drawback is the lack of accurate and comparable information that can be used by the banking industry.

Starting in 1996, the International Organization for Standardization (ISO) issued a series of comprehensive guidelines for incorporating environmental protection and pollution prevention objectives into industrial activity worldwide, known collectively as ISO 14000 (ISO, 1996). But how does the banking industry use environmental information in their credit extension and investment decisions?

The United Nations Environment Program (UNEP) has identified several types of environmental risks facing the banking industry (Vaughan, 1996), as has Rutherford (1994). These environmental risks are classified in Table 16.1.

Because banking operations by themselves are not highly pollution-intensive, pollution from their own operations is not the primary environmental concern of most banks. Their focus is on derived environmental liability through debt and equity transactions and derivative exposure through foreclosure and temporary asset management responsibility.

In addition, banks increasingly recognize that poor environmental practices by bank customers may reduce the value of collateralized property or increase the likelihood of fines or legal liability that reduce a debtor's ability to make payments to the bank. Besides the potential for suffering losses indirectly upon the contamination of collateralized property, banks in recent years occasionally have been held directly liable for actions occurring on properties in which they held a secured interest. Most noteworthy are cases like the Fleet Factors case in 1990,

Table 16.1 Potential Environmental Risk for Banks

1. Liability from the banks' own operations
2. Commercial lending and credit extension (debt) risks
   a. Reduced value of collateralized property
      → Cost of cleanup is capitalized into property value
      → Property transactions may be prohibited until cleanup occurs
   b. Potential lender liability
      → Cleanup of contamination on collateralized property in which the bank takes an interest
      → Personal injuries
      → Property damages
   c. Risk of loan default by debtors
      → Cash flow problems due to cleanup costs or other environmental liabilities
      → Reduced priority of repayment under bankruptcy
3. Investment (equity) risks
   → Effect of environmental liabilities on value of companies in which investment banks or their clients own equity
   → Upstream liability if the bank is a principal or general partner or owner

where the bank (Fleet Factors Corporation) was held liable for environmental damages incurred in the foreclosure process by a firm they hired to auction off assets [*U.S. v. Fleet Factors Corp.*, 901 F.2d 1550 (11th Cir. 1990), cert. Denied, 111 S. Ct 752 (1991)].

One notable case where a lender was held liable was the Mirabile case of 1985. In the Mirabile case, Mellon Bank was deemed sufficiently involved in day-to-day operations of the contaminated property that they were ruled not exempt under the secured interest provisions of CERCLA. Sometimes even the passive, temporary holding of property can put a bank in a risk position. In the Maryland Bank and Trust (MBT) case of 1986, the court held that MBT was liable for site cleanup under Superfund by simply holding the mortgage title for four years. The court deemed that MBT was "in a position to" uncover and resolve potential environmental problems at their secured properties. As a testament to the inconsistency of court rulings, the opposite was found in the Bergsoe Metal Corp. case of 1991, wherein the courts ruled that the lender could not be held liable unless actively participating in the management of the site. The court-generated confusion led the EPA to establish lender-liability rules in the hope of clarifying liability conditions.

Putting the details of the legal debate aside, there is increasing evidence of financial risk associated with poor environmental performance. Various studies have found a positive correlation between environmental performance and financial performance (Hamilton, 1995; Hart, 1995;

Blacconiere and Patten, 1993). So from one perspective or another, investment banks must consider environmental performance in deciding whether to invest in companies or advising clients to do so.

The ability of banking credit and investment practices to reflect environmental factors depends on a bank's ability to obtain and use accurate and reliable environmental information. However, such information is difficult to gather and even more difficult for the typical banker to interpret. ISO 14000 and sustainable development reports provide an opportunity for expanding the information base of the environmental performance of industrial entities and packaging it in a more user-friendly fashion. ISO 14000 is a series of voluntary compliance standards for environmental practices. These reports and other (SEC/EPA) ASTM standards being established are slowly establishing a consensus across a broad consortium of governments, businesses, and standardization organizations throughout the world regarding environmental performance.

Environmental management system standards such as ISO 140001 are being structured to be applicable to virtually any industrial producer. They cover:

- The establishment of an environmental policy;
- Environmental planning;
- Policy implementation and operation;
- Monitoring and corrective action programs; and
- Management review.

In so doing, the standard lays out a foundation for improving environmental performance through the establishment of environmental goals, implementation plans, monitoring programs, and corrective action programs.

It is important to understand the distinction between debt and equity transactions. Debt/equity distinction is a useful categorization of banking transactions. With regard to the environmental policy, vigilant environmental due diligence is advisable.

Practices for the Commercial Banking Community

There are several guidelines, standards, and regulations to help lenders limit their environmental liability. Several federal statutes increase the potential environmental liability for banking institutions—RCRA, CERCLA, TSCA, CWA, and CASA. Probably the most critical is the Comprehensive Environmental Response Compensation and Liability Act (CERCLA), enacted in 1980. This act established a wide net of parties potentially liable for the costs for remediating contaminated property, including increased environmental liability for banking institutions (FDIC, 1993).

Guidance is available via several federal regulatory agencies:

- The Office of Thrift Supervision (OTS);
- The Federal Reserve (The Fed);
- The Office of the Comptroller of Currency (OCC); and
- The Federal Deposit Insurance Corporation (FDIC).

The OTS has published several guidelines for lending institutions on environmental risk. They include the following:

- The 1989 issuance of Thrift Bulletin 1, which was entitled, "Environmental Risk and Liability: Guidelines on Development of Protective Policies and Reporting";
- The 1991 issuance of "Environmental Assessment Requirements for Properties Securing Loans Insured by Fannie Mae"; and
- The 1994 issuance of "Environmental Hazards Management Procedures."

The Federal Reserve has published guidelines on environmental liability for banks, entitled "Environmental Liability" (Federal Reserve, 1991). The OCC has issued guidelines for nationally chartered banks: "Banking Bulletin 92-38" (Ward, 1996) basically recommends that nationally chartered banks protect themselves from environmental liability by not participating in the management of properties for which they have a secured interest.

FDIC guidelines are considered to be the most comprehensive of the group. The FDIC recommends that banking institutions assess the potential adverse effects of environmental contamination on the value of real property and the potential environmental liability associated with the real property (Ward, 1996). It also suggests tailoring the environmental management and review process to reflect the type of lending an institution does and include securing approval by the bank's board of directors. It stresses that the lending institution must carefully follow these policies throughout the loan origination, renewal, refinancing, workout, and pre- and post-foreclosure stages. It cautions that the FDIC will look unfavorably upon a lending institution's failure to comply with these guidelines. Before a real property loan is made, the FDIC recommends conducting an initial environmental risk analysis (FDIC, 1993). This analysis consists of:

- A questionnaire and disclosure statement to be completed by the customer;
- An appropriate database search to determine whether the site or adjacent sites are Superfund sites, state cleanup sites, or other known environmental problem sites; and
- A field survey (with photographs) performed by trained personnel (Ward, 1996).

For foreclosures and trust transactions, the FDIC recommends that the bank evaluate the potential environmental costs and liabilities associated with taking title to the property (FDIC, 1993). There is a minimum ASTM standard for Phase I Environmental Site Assessments, but the scope of work for the Phase II Environmental Site Test is generally subject to the firm's environmental management's discretion and is based on the site's environmental history and general locality. Phase II may be omitted in some real property transactions, but it is generally performed for all foreclosures of commercial or industrial property.

ASTM has standards for conducting environmental site assessments for commercial real estate. The ASTM standard for conducting an environmental site assessment for commercial real estate was developed with respect to the range of contaminants within the scope of CERCLA. The ASTM standard was designed for property transfers rather than specifically for the banking industry. It fails to cover several issues desired by banks, such as regulatory requirements and compliance.

An environmental site assessment comprises the following four components:

- Records review;
- Site reconnaissance;
- Interviews with current owners and occupants of the property and interviews with local government officials; and
- The report.

The records review obtains and reviews records that will help identify recognized environmental conditions associated with the property. Site reconnaissance focuses on identifying recognized environmental conditions in connection with the property (e.g., stressed vegetation, starved soils, or surface water conditions). Interviews are geared toward collecting information that will indicate recognized environmental conditions in connection with the property. The report itself documents the analysis, opinions, and conclusions found in the assessment.

EPA Lender Liability Rule

In 1992, the EPA promulgated a lender-liability rule (National Contingency Plan, 40 C.F.R. Part 300, Subpart 1). The rule clarified that lenders would be protected from environmental liabilities under CERCLA as long as they adhered to certain basic rules (Scranton, 1992). Two years later, the rule was voided because the courts determined it exceeded the EPA's statutory authority [Kelly vs. EPA, 15 F.3d 1100 (D.C. Cir. 1994)]. EPA's lender-liability rule was reinstituted by legislation signed into law in 1996 as the "Asset Conservation, Lender Liability, and Deposit Insurance Protection Act of 1996" (in Title II, Subtitle E, Sections 2501–2505, P.L. 104–208).

The Lender Liability Rule defines a lender's liability relating to contamination involving vessels or facilities they finance. Lender liability differs between traditional lender actions (actions taken to protect a security interest) and acts of ownership, operation, or investment. The Lender Liability Rule defines participation in management as either exercising decision-making control over the borrower's environmental compliance or disposal activities or exercising executive or operational control, as opposed to exercising control over financial or merely administrative matters. As to the action of foreclosure on a contaminated property, the lender is protected under the rule as long as the property is sold in a commercially reasonable manner.

Post-Commitment Practices for Debt Transactions

Post-transaction monitoring is an important feature of the ideal environmental risk management program (Rutherford, 1994). Unfortunately, often much less attention is typically given to environmental issues after the financial institution actually commits funds, and typically ongoing environmental monitoring of a loan—while an accepted practice among non-U.S. banks—is not as widely accepted within the U.S. banking community. Bankers are focused primarily on the risk-avoidance side solely and not looking for the return on revenue opportunities to be found in innovative environmental practices (Environmental and Finance Research Enterprise, 1994).

After environmental site assessments and screening criteria, contractual covenants are the primary tool used by banks for managing and controlling environmental risk. Banks often use a "trigger" loan amount for requiring an environmental site assessment (Goodman and Hurst, 1995). Environmental risk management programs at banks typically involve the cooperation of account managers, risk managers, and bank customers. The goals of these programs are to identify environmental risks, assess the risks, and manage the bank's exposure.

A multi-step commercial lending process should be followed to incorporate environmental factors that are broken into:

- Preparation;
- Site inspection;
- Environmental document review; and
- A management system review.

The initial step is preparation, when account managers establish the nature of the environmental evaluation and the cooperation required of the customer. The next step consists of a site inspection and an environmental document review, which involve identifying environmental risks and liabilities and confirming the bank understands the customer and their problems and needs. During the third step, the management system review of the commercial lending process covers several topics, including environmental policies and procedures, resources, training and support, environmental liability, the compliance record, legal actions, audit programs, and any preventative actions the customer has initiated.

The later steps of the lending process involve analysis, loan structuring, credit approval, credit review, and loan management. If environmental risk is present, the bank's risk-management experts may be called in during the initial proceedings of those later steps. The actual loan structuring should vary for those companies with environmental risk versus those without the threat of environmental risk. For example, the bank may set up a trust to cover emergency and planned closure costs as well as post-closure costs. Furthermore, annual audits and quarterly environmental compliance certificates may also be required. Loan officers also have the option of asking for indemnification, insurance, and the submission of policies and procedures, as well. Loan management itself should vary for those companies thought to be environmentally risky. Their environmental matters will require regular review and monitoring.

Whereas environmental diligence is now a standard part of most real estate secured debt transactions, environmental due diligence is still relatively new to the credit process for equipment financing, general lines of credit, project finance activity, and other forms of credit. The challenge for banking institutions is to identify a consistent method to quantify the environmental issues to allow for integration into the core model. As it stands, the actual environmental due-diligence review is increasingly part of the basic credit process but not necessarily part of the evaluation model or score itself. While lagging the North American banking community on the environmental real estate due-diligence front, the European banking community appears to be significantly ahead of the Americans when it comes to evaluating environmental financial issues.

Besides CERCLA liability, there are other forms of lender liability imposed by other federal and state environmental statutes, worker safety standards (e.g., OSHA), or through third party lawsuits. There are also other factors that merit a bank's attention to environmental factors, such as the protection of collateral and the risk of borrower default that are not directly affected by the lender liability legislation. ISO 14001 increasingly serves a meaningful role in helping issuers of debt evaluate environmental risk on a pre-commitment and, to a lesser extent, post-commitment monitoring basis. ISO 14001 compliance (or noncompliance) information should be integrated into current environmental due-diligence processes on any form of credit as it pertains to any plant or equipment extension of credit.

ISO 14000 offers bankers consistent and comparable data that allows them to compare similar types of financing transactions. With the introduction of ISO 14001 and the development of an information framework (e.g., a sequence of questions) tied to it, credit officers can compare firms and plants on each facility's specific approaches to environmental management systems and their perception on how these differences in practices will affect relative risks. However, success in using ISO 14001 will depend on the extent to which customers see a connection increasingly recognized by the banking community. Although historically, equity-related bankers have generally been less attentive to environmental factors than bankers concerned with debt transactions (commercial bankers).

Practices for the Equity Banking Community

Relative to equity banking, the Securities and Exchange Commission (SEC) requires registrant companies to disclose environmental liabilities through their SEC filings. The disclosure requirements revolve around:

- Capital expenditures;
- Compliance policy;
- Litigation; and
- Additional information so that disclosure is not misleading.

The intent is to allow investors access to information on any impending environmental liabilities. The court case that spurred the SEC action also acknowledged the desire of so-called "ethical" investors to invest in companies concerned about the environment (*NRDC v. SEC*, 432 F. Supp 1190, D.D.C 1977). Reporting requirements are generally subject to the qualification that the expenditure or liability must be "material." In the past, this has opened up the potential for differences in interpretation. More to the point, historical "materiality" has been applied on a discrete basis to individual environmental issues within a company and not on an aggregate basis. However, the SEC appears to be moving away from the "individual" materiality approach to a more "aggregate" analysis.

When environmental issues are looked at more closely within the investment banking community, they tend to be viewed as either "deal killers" or as acceptable risks. Ironically, the ISO 14001 process should be of greater value to equity investors because the absolute level of risk is greater for investors undertaking a potentially large equity stake in a company versus a tender extending funds to a borrower. CERCLA liability aside, the lender's exposure is more-or-less limited to the funds extended, whereas the entire equity stake could be at risk for an investor.

"Salomon Smith Barney has been a leader on Wall Street in hearing the message that sound environmental management practice by a firm is a signal of future financial success for that firm." Lisa Leff, Director, a money manager in the Social Investment Program at Smith Barney Asset Management, said that "the amount of money entering 'socially enlightened' funds is much larger than most people would think—currently, an estimated 10% of all investment, or a total of $1.2 trillion, is connected to social investment. This segment is continually expanding." The mounting costs of environmental compliance and recognition that cost savings relating to pollution prevention and environmental sustainability are now being widely achieved.

In the past, the industry approach to environmental management practice has been to meet regulatory requirements. This has led to a tendency toward implementing only end-of-pipe solutions to address their environmental problems. However, some unique "role model" examples are emerging regarding the benefits of a more proactive environmental management approach, and the results of these efforts are beginning to be quantified. The pharmaceutical giant Merck began investing heavily in pollution prevention efforts a decade ago. Initially, it was felt that this would be pure cost for the company. Merck was pleased to find that these environmental initiatives resulted in significant cost savings. Similarly, Baxter Corporation, in its 1994 annual report, identified $23.4 million in new profits due to environmental programs. In addition, 3M, through its Pollution Prevention Pays (3 Ps) initiatives, has realized over $50 million in savings in a given year on a consistent basis. IBM states that it has had a $2-to-$1 return on its environmental investments.

But environmental is more than just "up-the-stack" industries. An example of imaginative environmental initiatives is the carpet company, Interface, Inc. Interface decided to lease carpet services rather than selling carpet as a product, while maintaining the highest quality standards for their customers. By taking back used carpeting and recycling it, Interface closed the product life-cycle loop. Using an innovative carpet squares product design approach, the company was able to implement a cradle-to-grave accountability for the products it uses in providing its services. Revenues have grown exponentially and profits have followed. Hopefully, this type of innovative responsible-environmental-management approach has the potential to become normal operating procedure in the future for manufacturing firms.

Academic studies (special issue of *Journal of Investing* for Winter 1997) show positive relationships between sound environmental practice and rising stock prices; not a single study has demonstrated a negative correlation. Recognize that while a firm's financial performance is conveyed to investors in company annual reports and SEC filings, there is no similar, clear mechanism for the reporting of environmental performance. Over 200 U.S. companies do issue environmental reports, often based on a "code of conduct" on environmental performance. The wide spectrum of reporting approaches make meaningful comparisons difficult, plus the existing environmental reporting tends to aggregate all environmental activities over a long time period, making it difficult to discern the effectiveness of any given environmental project or establish trends over a number of years.

Besides the SEC Environmental Reporting Requirement, there are required reports to the government, such as the Toxic Release Inventory (TRI), that can be used by the banking community to evaluate financial proposals. TRI lists the amounts of toxic materials emitted by a given firm. Unfortunately, TRI is self-reported, done on a site-by-site basis. The TRI data is not timely; it is usually more than a year late. As such, the data is more oriented toward capturing potential "past" consequences" and not toward "future" consequences of operational practices. Thus, the environmental information that is available is "reprocessed" by the EPA; so it is reported by "site," not by firm. Plus, be aware that even aggregated data of hazardous waste generation, air and water emissions, chemical spills, and Superfund liabilities give only minimal insight into the economic impact of the firm's environmental decisions. Problems with the TRI start with its historical focus. Measurement is based on past emissions with no estimate of likely future performance. Thus, the TRI numbers may not reflect current management philosophy or approaches.

Another difficulty for investors in evaluating environmental management is that often Financial and Investor Relations (IR) personnel are often left out of the "environmental loop." EHS recommendations are implemented in Operations and their outcomes are reported to the CEO with minimal communication to IR. A break in communication can also result from language differences between EHS and the IR. The IR people and the CFO generally view environmental management as a "cost," not as an area for potential profit and competitive advantage. Because the IR and the CFO are responsible for disseminating financial information, environmental management success is often neglected in communications with Wall Street analysts.

Recognize too that security analysts suffer from tremendous time pressure to produce opinions daily and quarterly, so they value information that is cheap, fast, and accurate. As a result, the average sell-side analyst, portfolio manager, or investment banker does not typically read corporate environmental reports, but views environmental issues in the "risk-compliance–cost" framework to the investor's potential risk. For example, an analyst holds a conversation in confidence with a Midwest-based utility, in which the utility claimed it had decided to ignore current EPA regulations for air emissions because of the costly nature of the required pollution control equipment. The company instead planned to upgrade equipment according to its standard capital expenditure cycle (10–15 years out), making the assumption that regulators 'wouldn't notice,' and that regulations would be significantly different at that point in the future anyway. When told of this decision, the analyst and his peers saw it as 'good news' for the stock. It meant the utility would avoid capital costs in the short term, increasing its reported income. The analyst was not worried that the long-term (two years or more) prospects of the utility were shaky, especially if the EPA should uncover the impropriety, because he could bail out of the security at any time. Thus, the utility company was rewarded with favorable reviews in the short run at the cost of potentially increasing its environmental practices in the longer term in a dramatic fashion.

To counteract this short-term philosophy, social- and environmental-related indices are developing. Kinder, Lydenberg & Domini developed the Domini Social Index that has inputs relating to the social responsibility and the quality of the environmental management of firms. Innovest and Elipson have also designed models based on proprietary screens for environmental management practices. The EPA is also looking into starting its own environmentally screened index. As a proxy thumbnail indicator of the quality of a firm's environmental focus apart from environmental management indices, investors can also use a firm's accounting practices. For example, an indicator of rudimentary environmental awareness may be waste, treated as a cost and aggregated into overhead. An indicator of moderate environmental awareness identifying the cost of waste and other environmental matters by the department where the product is produced, or to the product itself, can be an indicator of moderate environmental awareness. Establishing cost, including the cost of transportation to disposal and of "value added" prior to disposal, to the product produced can be an indicator of high environmental awareness.

Finally, everyone should recognize that individual investors can use "shareholder resolutions" to bring environmental issues to the attention of the public and other shareholders despite corporation and management postures to the company. It takes only $1000 of common stock held for one year to file a resolution to include such issues on the annual shareholder proxy and on the agenda of a firm's annual meeting. As little as 3% of a proxy vote can prompt significant investor concern about corporate environmental management behavior. In short, shareholder advocacy is the type of action that gets management's attention to explore potential deficiencies in their policies, but it is a risk that should not occur in a pro-active environmental management setting.

Integrating Environmental and Financial Performance

Economic drivers can powerfully enhance or impede the achievement of corporate environmental goals. Growing population and consumption places increasing demands on limited natural resources and has increased citizen and government concern.

Establishing environmental business value requires better communication between a company's technical and environmental side and its financial and business managers to establish the ways environmental consider-ations integrate with core business strategies, such as:

- Enhancing the business franchise;
- Improving process efficiency;
- New product development; and
- Building new markets.

Measuring the results of these strategic environmental initiatives involves establishing how a company is valued, what environmental measures its investors use, how the aspects of its environmental strategies affect those measures, and how to collect data that demonstrates success. This calls for using measurements and terms that are understood and used by the financial community. It requires expressing results in terms of business goals by:

- Communicating the data in financial terms that are relevant to the concerns of investors;
- Ensuring that all company officials provide a consistent message;
- Providing credible information; and
- Reporting in a manner that is comparable to other companies in a given sector.

A growing segment of the financial community is recognizing and rewarding the publicly traded businesses that are strengthening busi-ness-environmental linkages. In addition, insurance companies and lend-ers are increasingly recognizing the business-environmental linkages by selectively adjusting their rates based on environmental criteria. In the end, the business-environmental linkage really deals with familiar ques-tions: quality of management, risk exposure, brand image and reputation, overall operating efficiency, growth, market access, and particularly pension fund management.

Environmental issues are increasingly connected with economic devel-opment, trade, and the global demand for goods and services. Beginning with the World Commission on Environment and Development in 1987, there has been an increased attention on the concept of sustainable devel-opment, stressing this linkage and the need for an integrated approach to business and the environment. At the United Nations "Earth Summit" in Rio de Janeiro in 1992, business engaged as a major contributor toward solutions, and as a result the dialogue about the interrelatedness of economic development, environmental protection, and social welfare expanded. The new approaches take advantage of corporate knowledge about how to optimize their products, processes, and business strategies from an environmental sense.

The globalization of the economy increases the importance of access to high-growth international markets. A company's environmental record and reputation can hamper this new market entrance. Also from a broad image perspective, transnationals face greater responsibility for environmental impact along their entire value chain, including suppliers. Companies are increasingly selling a service rather than a product, allowing their customers to be less commodity or capital intensive. This creates the opportunity to capture environmental benefits through reuse and recycling, and in so doing "locking in" customers and creating a competitive edge for the supplier. Product differentiation, market position, and business advantage are increasingly based on environmental perceptions of products and services.

The environment can make significant contributions to core business strategies in several ways:

- Protecting the business franchise by reducing risk and opening new markets;
- Changing processes to improve efficiency, thereby increasing margins and return on investment;
- Fostering product change to increase competitive advantage; and
- Building new markets that reposition the company in the marketplace.

To increase economic value from environmental positioning as part of a business strategy, companies must understand how their environmental activities contribute to one or more of these strategies and to communicate this information effectively throughout the organization. This may require new lines of communications. Many of the existing environmental reports are anecdotal or include data on the aggregate levels of pollutants produced but with little information on their financial impact. Unfortunately, voluntary, non-standardized, and typically unaudited reports and questionnaires are rarely useful to financial analysts. To effectively convey the financial value of its environmental strategies, the environmental reporting emphasis shall vary depending on the industry. A company must establish how it is being valued in the marketplace and provide analysts and investors with information that relates to the means of valuation being used. For example, companies in an industry that cares primarily about the ability of a company to introduce new products should explain how their environmental management program provides a source of insight for new product development versus stressing environmental cost savings that might be more applicable to other industry settings.

As to environmental cost savings, investors beware! Many environmental strategies being pursued by companies are process-changing in nature. These often lead to cost efficiencies and increased earnings. However, simply reporting the savings achieved as a result of these initiatives can be confusing and potentially deceiving unless it includes the costs of implementing the measures. The net savings and not the gross savings are what contribute to earnings. Likewise, the persistence of savings and assumptions about how these savings apply to growing or shrinking lines of business are valuable to analysts.

	Franchise Protection	Process Changes	Product Changes	New Market Development
Business Value	Right to operate	Cost and liability reduction	Market share and pricing power through customer loyalty and reputation	New markets
Focus	Compliance	Efficiency	Value chain	Innovation
Main Financial Impact	Reduces earnings Reduces risks Can open new markets	Increases margins Reduces risks Often uses less capital, increases return on equity	Increases competitive advantage	Increases sales Increases competitive advantage
Barriers to Integration into Financial Analysis	Risk is not an explicit variable in most valuation models	Many diverse sources of small earnings improvements Risk is often not explicit variable	Quantification of competitive advantage difficult	Quantification of competitive advantage difficult

Exhibit 68. Environmental strategies: a corporate view.

The critical issue for gaining recognition of environmental business value is the communication in terms that are understood and valued within the financial community. This requires an explicit understanding of the various goals that can be achieved in part by environmental actions. Exhibit 68 identifies the links between environmental value strategies and the interests of financial analysts and investors. Most companies' environmental strategies should fall under one or more of the column headings; however, analysts' and investors' interest lies in the financial impacts of those strategies, summarized in the third row.

Analysts, money managers, and investors will pay greater attention to environmental issues when they are shown the connection between environmental strategies and margins, markets, and growth, and how these strategies can directly contribute to increased earnings and multiples. To improve the quality and quantity of financially relevant information on company environmental initiatives—to reshape the message—business and environmental managers must address five critical communication needs:

- *Effective internal measurement and communication.* Because company representatives are the primary source of information for analysts, building internal awareness and understanding is a pre-condition for closing the external communication gap.
- *Relevant information.* The company should target environmental communications to the following issues of concern to financial analysis:
    - Financial performance—environmental terms, such as tons of emissions or tons of waste, must be translated into financially relevant terms.
    - Business functions—communications should demonstrate the potential driving impact environmental initiatives have over the company decision-making as a whole.
    - Business strategy—environmental strategies should be communicated in the context of product, market, technology, or cost-reduction opportunities to be gained through environmental innovation.
- *Consistent information.* Analysts and investors take their cues from meetings with corporate senior management—particularly, the chief financial officer and investor relations manager. Yet these managers often discuss the environmental issues as risks and liabilities. All senior managers should be equipped to discuss the environment as a strategic management issue. They must be able to articulate how the company's approach is allowing the firm to exploit competitive opportunities.
- *Credible information.* However, the company and senior management must—at all costs—avoid the trap of "greenwash." Environmental claims and projections should be examined critically before cited publicly to avoid a loss of confidence in management and the company.
- *Comparable information.* Analysts must be able to compare data across firms. Multi-stakeholder initiatives (discussed elsewhere in more detail) are designed to help define and disclose relevant, comparable environmental performance information. Initiatives like the CERES Global Reporting Initiative are critical.

However, in general financial institutions lagged behind corporate community leaders in recognizing and attempting to value environmentally driven corporate change. The majority of investment firms have a good grasp of environmental liabilities. However, many investment professionals are still reluctant to make the connection between "beyond compliance" environmental performance and shareholder value creation. Their attitude is supported by often inadequate accounting practices and data management systems that fail to quantify the financial value of environmental actions.

Among the financial community's early adopters of environmental value to investments were the *socially responsible investment* (SRI) practitioners. SRI groups invest in a way that is consistent with certain social and ethical concerns. As a result, SRI portfolios typically screen to include or exclude securities of certain types of companies. Unfortunately, traditionally SRI has been relatively unsophisticated in its analysis of company practices apart from qualitative "social screening." However, in recent years environmental screening has evolved from simple avoidance of high-impact industries (e.g., petroleum, forest products, chemicals) to analyzing companies on their own environmental merits. As such, there is an increased focus on a "best of class" approach. Recent SRI analysts place an increasingly heavy interest in a company's environmental initiatives. To that end, the trend toward company environmental reporting is a response to this demand. To further complicate matters, traditionally most social investment funds have had a split (in terms of both methodology and staff) between social and environmental research. However, recent sustainable development reporting combines these issues, generating a need for quantifying the links between social, environmental, and financial performance. Recognize too that SRI is not a fringe phenomenon. SRI is now increasingly demanded by foundations, pension funds, schools, labor unions, health care agencies, and a much wider spectrum of individuals, both in the United States and abroad.

In recent years, a more focused approach to integrating environmental and financial analysis emerged in Europe in the form of "environmental value" funds. These funds are constructed by integrating traditional investment analysis with an analysis to identify the top environmental performers in each target industry section. The benefit of this "best of class" approach does not exclude entire industry sectors. In addition, it also explicitly acknowledges a positive, identifiable link between what is good for the environment and what is good for shareholders.

Key factors in an "environmental value" fund approach include a range of quantitative indicators relating to climate change, ozone depletion, toxic release, intensity of water and energy use, environmental liabilities, and environmental management quality. As discussed earlier, a number of firms (e.g., Canadian-based Innovest and Sustainable Systems Associates, and the Swiss firm Bank Sarasin) have done substantial work designing new environmentally based analytical tools to help companies and portfolio managers uncover the "hidden" value potential from strategic environmental management, thus allowing investment analysis to identify companies with superior appreciation potential. In general, each model attempts to balance a company's level of risk with its capacity to control that risk and to capitalize on environment-driven opportunities. On the risk side, the key factors being examined include:

- Historical contingent liabilities;
- Operating risk exposure arising from such factors as toxic releases, hazardous waste disposal, and product risk liabilities; and
- Eco-efficiency and sustainability risk, arising from such factors as energy intensity, raw material usage, the durability and recyclability of a company's products, and a company's exposure to consumer value shifts.

In contrast to investment firms, the key issue for commercial lenders is how environmental performance influences the borrower's ability to generate sufficient cash flows to repay the loan. Some lenders do not share in the upside gains realized by a business; their primary interest is in the downside, e.g., the risk. Consequently, their focus has been on the nature and extent of a borrower's environmental liabilities, capital expenditures, and the operating costs required to meet existing and anticipated laws and regulations. They are also concerned with any exposure to litigation as a consequence of concerns arising from a company's products or processes, past, present, or future—but unfortunately with emphasis on past and present. However, a small but growing number of lending institutions led by the major Swiss and German banks believe that a customer's eco-efficiency and the quality of its environmental management system are as important as environmental risk in today's increasingly sustainability-driven marketplace. These lenders believe that environmental performance is a strong indicator for overall corporate risk and are not only using environmental issues to decide whether a loan should be made but to even calculate the risk premium, at least in some industry sectors.

Insurers are also increasingly using environmental criteria as factors in their policy negotiations with clients—in some cases, offering discounts of up to 30% on premiums for environmental impairment liability insurance based on the verifiable degree of success of their program. As this trend grows, particularly in industries with relatively high insurance costs, the impact on costs and earnings is likely to be increasingly noted by analysts and investors and reflected in the share price. In summary, environmental initiatives offer the analyst and investor a new tool to assess a company's potential for success.

An emerging view of fiduciary responsibility may also be a factor that focuses greater attention to the business value of the environment. In the past, many pension fund managers and other fiduciaries have refrained from looking at environmental and social issues on the grounds they believed that they could not consider these ethical or moral concerns because their legal responsibility is to maximize the return on their funds. This view of fiduciary responsibility has begun to change. In recent years, the U.S. Department of Labor has issued an opinion that a "socially responsible" mutual fund would not necessarily be inconsistent with fiduciary standards under ERISA. As fund managers pay more attention to the positive results of considering environmental factors, a real opportunity exists for leading companies. This will begin to turn the fiduciary responsibility from a prohibition on considering environmental factors into a requirement. Pension fund managers will have to consider these issues to meet their legal obligations.

Environmental issues such as global climate change, endocrine disruptors, biotechnology, and others are resulting in profound structural change in some industry sectors. Ignoring environmental drivers could mean missing an important element of competitive advantage, both in a company's planning and in its assessment by analysts and investors.

References

ASTM. "Standard Practice for Environmental Site Assessments: Phase I Environmental Site Assessment Process." (Philadelphia, PA: ASTM, 1993.)

Bankers Roundtable. "Roundtable Survey: Environmental Liability of Secured Parties and Fiduciaries." (Washington, DC: Bankers Roundtable, 1995.)

Bennett, Mark J. "Environmental Risk Policies at Financial Institutions." *The Journal of Commercial Lending*, October:45–50, 1994.

Bisset, Doug. "Managing Environmental Risk: A New Responsibility for Banks." *The Bankers Magazine* 178(2):55–59, 1995.

Blacconiere, Walter, and Dennis Patten. "Environmental Disclosures, Regulatory Costs, and Changes in Firm Value." *Journal of Accounting and Economics* (December, 1993).

Blumberg, Jerald, Åge Korsvold, and Georges Blum. "Environmental Performance and Shareholder Value," World Business Council for Sustainable Development, 1997.

Brown, Johnine J.. "Progress in Reducing Lender LiabilityUnder Environmental Law." *The Bankers Magazine* 179(4):30–33, 1996.

CEEM Information Services (CEEM). *ISO 14000 Questions and Answers.* (Fairfax, VA: CEEM Information Services, 1996.)

Descano, Linda, and Bradford S. Gentry. "Communicating Environmental Performance to the Capital Markets," *Corporate Environmental Strategy*, Spring, 1998, 14–19.

Descano, Linda, and Bradford S. Gentry. "How to Communicate Environmental Performance to the Capital Markets," *CMA News*, April 1998, 33–37.

Ellis, Billie J., Jr., Sharon S. Millians, and Sandra Y. Bodeau.. "Helping a Lender Develop an Environmental Risk Program." *The Practical Real Estate Lawyer* July:81–94, 1992.

Environment and Finance Research Enterprise.. "Global Survey on Environmental Policies and Practices of the Financial Services Industry." (Chapel Hill, NC: Environment and Finance Research Enterprise, 1994).

Environmental Grantmakers Association. *Philanthropy as Stewardship: Recommended Principles & Practices for Operating in an Environmentally Responsible Manner.* 3d ed., 1997.

FDIC. "Environmental Liability: FDIC Guidelines on Risk Prevention Programs." Financial Institution Letter FIL-14-93. (Washington, DC: FDIC, 1993).

Federal Reserve. "Environmental Liability." Discussion paper for external distribution. Division of Banking Supervision and Regulations, 1991.

Ganzi, John, Frances Seymour, and Sandy Buffett. *Leverage for the Environment: A Guide to the Private Financial Services Industry.* (World Resources Institute, 1998).

Gentry, Bradford S. and Lisa O. Fernandez, *Valuing the Environment: How Fortune 500 CFOs and Analysts Ensure Corporate Value*, United Nations Development Programme, 1997.

Goodman, W. Robert, and James R. Hurst. "The Perceived Impacts of Lender Environmental Liability on Alabama Banks." *The Southern Business and Economic Journal* 18(2):124–134, 1995.

Hamilton, James T. "Pollution as News: Media and Stock Market Reactions to the Toxics Release Inventory Data." *Journal of Environmental Economics and Management*, 28:98–113, 1995.

Hart, Stuart. "Does It Pay to Be Green? An Empirical Examination of the Relationship Between Emissions Reduction and Firm Performance." *Business Strategy and the Environment* (September, 1995).

Hart, Stuart. "Beyond Greening: Goal and Strategies for a Sustainable World," *Harvard Business Review* Jan–Feb 1997, 67–76.

ISO. "Environmental Management Systems—Specification with Guidance for Use." Reference number ISO 14001:1996(E). (West Conshohocken, PA: ASTM, 1996.)

Kerr, Bill. Personal communication between Bill Kerr, Bank Examiner and Policy Analyst, Office of the Comptroller of Currency, and Brian Murray (Research Triangle Institute. November 13, 1996.)

Leff, Lisa. Director of Money Manager, the Social Investment Program, Smith Barney Asset Management, Salomon Smith Barney. "Environmental Excellence as a Signal of a Firm's Future Financial Performance." Connecting Business & the Environment, A Seminar Series on Industrial Ecology, Seminar #12.

Meloy, Michael M. "Disclosure of Environmental Liability in SEC Filings, Financial Statements, and Debt Instruments: An Introduction." *Villanova Environmental Law Journal* 5(2):315–321, 1994.

Muller, K., J. de Frutos, K-U. Schussler, and H. Haarbosch, "Environmental Reporting and Disclosures: The Financial Analyst's View," European Federation of Financial Analysts' Societies, 1994.

Muller, K., J. de Frutos, K-U. Schussler, H. Haarbosch, and M. Randel, "Eco-Efficiency and Financial Analysis: The Financial Analyst's View," European Federation of Financial Analysts' Societies, 1996.

Novitski, David E. "Ongoing SEC Disclosure Requirements." In *Environmental Problems in Financing and Securities Disclosure: Avoiding the Risks* 35–118. (Practising Law Institute, 1991).

O'Brien, James P. "New Lender Protections Against Federal Superfund Liability." Chapman and Cutler Environmental Insights Fall:1–10, 1996.

Porter, Michael E. and Claas van der Linde. "Green and Competitive: Ending the Stalemate," *Harvard Business Review* Sept–Oct 1995.

Practising Law Institute. Environmental *Problems in Financing and Securities Disclosure: Avoiding the Risks*, 1991.

Reed, Donald J., "Green Shareholder Value, Hype or Hit?" *Sustainable Enterprise Perspectives* (World Resources Institute, September 1998).

Robbins, Lorne, and Douglas M. Bisset. "The Role of Environmental Risk Management the Credit Process." *Journal of Commercial Lending* 76(10):18–25, 1994.

Rutherford, Michael. "At What Point Can Pollution Be Said to Cause Damage to the Environment?" *The Banker*, January:10–11, 1994.

Schmidheiny, Stephan, and Federico J.L. Zorraquin. *Financing Change: The Financial Community, Eco-Efficiency, and Sustainable Development*. (Cambridge: MA: The MIT Press, 1996).

Scranton, David F. "Issues in Lending... Understanding the New EPA Lender Liability Rule." *Journal of Commercial Lending* 74(11):6–16, 1992.

Sesit, Michael R. "Disclosure Fails to Meet Needs of Big Investors: Survey Shows Institutions Feel Company Reports Yield Insufficient Data." *Wall Street Journal*. November 4, 1996.

Uncovering Value: Integrating Environmental and Financial Performance. (Washington, D.C.: The Aspen Institute, 1998).

Vaughan, Scott (ed). "Greening Financial Markets. Geneva: United Nations Environmental Programme." As cited in Schmidheiny, Stephan, and Federico J.L. Zorraquin. 1996. *Financing Change: The Financial Community, Eco-Efficiency, and Sustainable Development.* (Cambridge: MA: The MIT Press, 1995).

Ward, Elizabeth H. "Environmental Risk Management: The Why and How." *The Banker's Magazine* 179(4):19–24, 1996.

Weiler, Edward, Brian C. Murray, Sheryl J. Kelly, and John T. Ganzi. "Review of Environmental Risk Management at Banking Institutions and Potential Relevance of ISO 14000: Working Paper," RTI Project Number 5774-4. (Washington, D.C.: EPA, April 1997).

Chapter 17
Global Warming

Global climate change is evolving. The regulation of greenhouse gas emissions is still in the formative stage; the potential for firms new to the market is high. Most companies are still on an evolving learning curve. Whereas the United States has currently withdrawn from the Kyoto Protocol, the issue of regulating CO_2 and other greenhouse gas (GHG) emissions as a contributor to global climate change is not going away. The "Clean Skies Initiative" does not regulate CO_2, but numerous other pieces of legislation that regulate CO_2 have been filed in Congress. Most notable of these is Senator Jeffords' "Clean Power Act" (S.556). The current national multi-pollutant debate is over a "three-pollutant" (NO_x, SO_x, and Hg) versus a "four-pollutant" (NO_x, SO_x, and Hg plus CO_2) bill. The potential passage of any bill at this point is unclear, but the issue will continue to be part of the national debate. The question of regulating CO_2 is probably not an "if" but a "when," and by extension, "how."

At the state level, many states have passed or are considering regulations that address CO_2 or GHG emissions. Massachusetts has adopted the first regulations in the country that cap CO_2 emissions and set emission rate limits. New Hampshire is close to passing a regulation, and California is in the process of setting up a greenhouse gas registry.

Internationally, the Kyoto Protocol is moving forward, even without U.S. participation. At least 55 nations, accounting for 55% of the 1990 GHG emissions, were needed to ratify Kyoto for it to enter into force. More than 55 nations have ratified Kyoto, but the percent of emissions represented by these nations was below 55% by 2003. The European Union and Japan both ratified Kyoto, and all eyes turned to Russia. Russia's ratification brought Kyoto into force even without U.S. participation. Because it brings the percent of emissions above 55%, passage of Kyoto means that a large multinational U.S. company doing business in a country that is a participant in the Kyoto Protocol—particularly an energy company—can expect some form of regulation of GHG emissions.

There are still other national regulations being put into place. In Europe, Denmark and the U.K. have emission-trading programs in place. Denmark passed the CO_2 Quota Act in 1999. It places a mandatory cap on CO_2 emissions from electricity producers. Participation in the U.K. program is voluntary (though encouraged by government tax incentives) and covers all industries and all six Kyoto gases. In March 2002, the U.K. auctioned off allowances that will be used in a trading program to help meet voluntary emission reductions.

Compliance issues for companies currently differ and will continue to differ from state-to-state and country-to-country. Hence, a large multi-national company with facilities in different states and throughout the world may have multiple regulatory issues that affect it. How those regulatory programs do or do not interact with each other is complex and will require constant vigilance. This creates a high level of uncertainty and thus a need for our services to—at a minimum—track these changes in the regulatory arena.

Existing Market and Potential Revenue

Much of the early work on global climate change came from the realm of "think tank" activity. These studies and white papers included initial demonstration projects and sample protocols for establishing baselines, emission reduction verification, certification, and monitoring as people tried to get a handle on how all this stuff is really going to work. For example, papers on how to conduct emission inventories and how to verify emission reductions were completing at the Pew Center for Global Climate Change. The State of California is setting up a GHG registry. The World Bank and its Prototype Carbon Fund (PCF), as well as other international organizations, are engaged in the debate as well. Even the U.S. Department of Energy (DOE) has directed significant research money to carbon sequestration studies.

Individual companies are also active in defining their positions in the emissions trading market. The incentive for companies to take early action on reducing GHG emissions is for them to "learn by doing" and to potentially realize substantial savings if the emission reductions made now are recognized in a future trading market. There are currently several pilot-trading markets in place. An example of this is in Canada at Ontario's Pilot Emissions Reduction Trading Project (PERT). Some companies, notably BP and Shell, have developed internal trading markets.

For example, Entergy has a $25 million Environmental Initiatives Fund to support CO_2 reduction projects. Internal improvement methods include power plant heat rate improvements, natural gas leakage reduction, SF_6 containment, high-efficiency transformers, and the use of alternative fuel vehicles. External projects include forestry projects, coal mine methane capture/utilization, and end-user efficiency improvements. A total of 80% will be from internal reductions, with 20% from external offsets, and they expect to spend around $500,000 per project. Entergy has set up project selection criteria for screening and selecting projects. Level 1 of the project selection criteria includes the credibility of reductions (Entergy uses independent third-party verification). Level 2 criteria include cost effectiveness, strategic value, and media/public relations value.

Brokers

A central element to the Kyoto Protocol and many other efforts to facilitate the reduction of GHG emissions is the use of emission trading as a means of achieving a cost-effective solution. The current international market for CO_2 trades is $50 million. The market is projected to grow to between $25 billion and $700 billion. Consequently, many brokers have an interest in seeing the market develop. Two firms actively pursuing the market are CO_2 (an arm of Cantor Fitzgerald) and Natsource. CO_2 has developed an associate relations program with auditing firms, consulting firms, and engineering companies. Both companies also run trading simulation workshops, which may be a way of getting more familiar with them and of meeting other potential clients.

Because there are currently no government programs in place that have issued tradable allowances of credits, the only market in GHG trading is for either non-verified or *verified emission reductions* (VER). VERs carry a higher market value because they have been subject to an independent audit or verification.

A large portion of the early activity related to global climate change has been with governmental organizations. The Kyoto Protocol will require Annex B countries (the 39 emission-capped countries) to prepare plans on how they will meet their obligations under Kyoto. Non-Annex B, or developing, countries such as Mexico will need to have in place governmental structures, such as environmental ministries, that can certify emission reduction projects carried out under Kyoto's Clean Development Mechanism (CDM).

Global Climate Profile

As part of the response to the emerging global climate issue, corporations are encouraged to develop global climate profiles. Exhibit 69 provides a lexicon of global climate change terminology. The following outline lays out an organizational approach to establish its global climate profile:

1. *Corporate Climate Change Profile.* This initial step involves assembling basic information available on the issue of climate change. The profile is a status report on what is happening with regulations at the state, national, and international level. Finally, the profile should provide a first-cut needs analysis/risk assessment of what the company's current or potential exposure is due to efforts to regulate GHGs.
2. *Emission Inventory and Baseline Development.* This step is furthering the development of a profile. It should quantify the status of a company's GHG emissions, and set up the protocols by which the company can begin to monitor its GHG emissions. The work should

---

◆ GHG — Greenhouse gas. Typically refers to the six gases identified in the Kyoto Accord.

◆ GWP — Global warming potential. The measure of a gas's "radiative forcing" or ability to trap heat in the atmosphere. Conference of parties. Supreme body of UNFCC.

◆ IPCC — Intergovernmental Panel on Climate Change. Created by UNEP and WMO in 1988.

◆ UNFCCC — United Nations Framework Convention on Climate Change — Established at the June 1992 Rio Earth Summit.

◆ Kyoto Protocol — Refers to agreements reached in December 1997 when signature countries agreed to levels of emission reductions (average 5.2% below 1990 level).

◆ Carbon Equivalent (CO_2e) — Measurement of the global warming potential of a greenhouse gas.

◆ Baseline — Point from which emission reductions are measured. May be static, adjusted, or benchmarked.

◆ Sequestration — The capture of storage of carbon through forestry, land or soil conservation, or CO_2 recovery and injection.

◆ Leakage — Apparent reductions that are achieved in one location, only to be generated in another.

◆ Additionality — Reductions in emissions must be in addition to what might have otherwise occurred.

◆ Banking — Ability to store and take credit for reductions prior to enactment of requirements.

Exhibit 69. Global climate change lexicon.

---

include a review of an existing or a recommendation for an emissions database. It should provide an assessment of emission boundaries and strategic advice on setting those boundaries.

3. *Mitigation Strategies.* Based on the corporate profile, emissions inventory, and baseline development, a mitigation strategy should be developed. The mitigation strategy should include both engineering and non-engineering alternatives to meeting or reducing GHG emissions. The product should include an outline with associated costs of providing emission reductions through each mechanism, such as GHG trading or engineering upgrades. It should also provide a list of offsite GHG reduction options similar to those recognized through Kyoto's projects.

4. *Long-Term Emission Monitoring.* The company should develop a protocol or work plan for monitoring its GHG emissions that is auditable for the future verification of emission reductions. This could include software and other information management systems to support monitoring and reporting at both the plant and fleet level.

Gas	GWP
✧ Carbon dioxide (CO_2)	1
✧ Methane (CH_4)*	21
✧ Nitrous oxide (N_2O)	310
HFC-23	11,700
HFC-125	2,800
✧ HFC-134a	1,300
HFC-143a	3,800
HFC-152a	140
HFC-227ea	2,900
HFC-2361a	6,300
HFC-4310mee	1,300
✧ CF_4	6,500
C_2F_5	9,200
C_4F_{10}	7,000
C_6F_{14}	7,400
✧ SF_6	23,900

*The methane GWP includes the direct effects and those indirect effects due to the production of tropospheric ozone and stratospheric water vapor. The indirect effect due to the production of CO_2 is not included.

Exhibit 70. Six Kyoto greenhouse gases (GHG).

5. *Third-Party Independent Verification and Certification of Emission Reductions.* For companies currently participating in GHG trading programs, third-party independent verification of the company's emission reductions should be attained so that emission reductions could then be traded in the GHG market.

Global warming potential (GWP) (Exhibit 70) is a measure of a gas's ability to trap heat in the atmosphere. GWP was developed by the Inter-governmental Panel on Climate Change (IPCC) in 1996 and is measured over a 100-year time horizon. The IPCC is laid out in Exhibit 70. GWP compares the ability of each GHG to trap heat in the atmosphere relative to other gases, with CO_2 used as the basis. For example, sulfur hexafluoride (SF_6), used in gas-insulated switch gears, is 23,900 times more effective at trapping heat in the atmosphere than CO_2.

Results indicate that GHGs are persistent over time. Also, the increase in GHGs causes an increase in radiant forcing, which leads to an associated increase in global temperature. The potential impacts of global temperature increase include:

- A rise in sea level, which could impact tens of millions of people in small island states and low-lying coastal delta regions;

- Regional changes in climatic events, i.e., drought, heat waves, floods, hurricanes; and
- The disruption of ecological systems.

Global Climate Summary

In short, GHGs are global pollutants, and the effect is not restricted to a regional or "downwind" area. Nor will the response to controls be immediate, given GHG's persistence in the atmosphere for hundreds of years.

The Kyoto Protocol adopted in 1997 by parties to the Convention on Climate Change in Kyoto, Japan, calls for binding emission limits to reduce GHG emission to 5% (on average) below 1990 levels for the period 2008 to 2012. The U.S. currently represents 25% of the global emissions. The methods that have been established to control GHGs are:

- Reduction in use
- Sequestration
- Emission trading

Carbon sequestration refers to mechanisms to capture and store carbon (sometimes referred to as a "carbon sink") in a manner that prevents it from being released into the atmosphere for a specified period of time. There are two basic methods of sequestration:

- Passive—forestry, land, and soil conservation methods; and
- Active—recovery of waste CO_2 and injection for storage.

Another carbon sequestration example is geologic sequestration, which includes:

- Oil and gas recovery—CO_2 pumped in for enhanced oil recovery (EOR) used 32 million tons of CO_2 in 1998; and
- Coal bed methane displacement through CO2 injection.

Emissions trading was pioneered by the 1990 Clean Air Act's (CAA) Title IV Acid Rain Program, which authorized trading of SO_2 allowances. NO_x allowance trading also occurs under CAA. The SO_2 and NO_x programs have shown that emission trading is an economically efficient way to achieve emission reductions. CO_2 emissions-trading programs are scattered and have developed on an ad hoc basis. As the Kyoto debate moved forward, multiple trading programs developed. Current trading is in VERs, not allowances or credits, which are government-created tradable commodities. Rules and protocols vary with each program, and some are still in the development phase. This fragmentation of programs affects market price, increases transaction costs, and reduces liquidity.

Three basic methods of emission trading proposed under Kyoto are:

- International Emission Trading (IET)—trading of assigned amount units (AAU) among "Annex B" countries.

- Joint Implementation (JI)—the creation of emission reduction units (ERU) by building and investing in emission reduction projects in "Annex B" countries.
- Clean Development Mechanism (CDM)—allows the creation of certified emissions reductions (CER) by "Annex B" countries through the building and investing in emission reduction projects in "Non-Annex B" countries.

Current trading programs include:

- The United Kingdom's voluntary program, with economic incentives that include all GHGs
- Denmark—a binding program for CO_2 emissions only, as well as Emissions Reduction Unit Procurement Tender (ERUPT)
- Shell and BP have internal trading programs
- U.S. Initiative on Joint Implementation (USIJI)
- International—Actions Initiated Jointly (AIJ)
- Canada—Pilot Emissions Reduction Trading Program (PERT)
- World Bank—Prototype Carbon Fund (PCF)

The CO_2 emissions trading market is potentially huge. Historically, the price has ranged between $1 and $3 per ton of CO_2 equivalent (CO_2e), but is projected as high as $15 to $50 per ton of CO_2e. An estimated 65 GHG trades equaling more than 1000 metric tons of CO_2e have occurred in the world since 1996. U.S. action on GHG emissions has occurred at both the federal and state levels. Federal actions include:

- Section 1605(b) of the Clean Air Act—the voluntary reporting of GHG emissions
- Senator Jeffords' "Clean Power Act" Bill (S.555)—which would amend CAA to include CO_2 reductions
- DOE Clean Coal/Carbon Sequestration Initiatives

State actions include that more than 20 states have considered or passed legislation related to GHGs. Despite all the heated rhetoric that flies around the global warming issue, business does take it seriously.

> "AEP accepts the views of most scientists that enough is known about the science and environmental impacts of global climate change for us to take actions to address its consequences."
>
> **—Dale E. Heydlauff, Senior Vice President-Environmental Affairs, American Electric Power, on May 23, 2001, before the Senate Subcommittee on Science, Technology, and Space**

Reasons to take action on global warming include:

- Early action is likely to reduce cost; and
- Companies gain a firm experience on the learning curve.

It also will allow leaders to help influence policy and regulatory developments and establish working partnerships with NGOs. The latter has a high public relations value and may allow a firm to be better able to manage perceived risks down the road.

References

"A report by the Competitive Enterprise Institute, a pro-business group of global warming skeptics." (http://www.cei.org/pdf/5430.pdf).

"China Approves Kyoto Protocol" (http://chinese-school.netfirms.com/news-article-China-Kyoto-Protocol.html).

"Climate Change 2001: Working Group I: The Scientific Basis, 3.7.3.3 SRES scenarios and their implications for future CO_2 concentration." (http://www.grida.no/climate/ipcc_ tar/wg1/123.htm).

"Final Report of U.S. Climate Change Science Program." (http://www.climatescience.gov/Library/sap/sap1-1/finalreport/default.htm).

"The Impact of the Kyoto Protocol on U.S. Economic Growth and Projected Budget Surpluses." (http://www.accf.org/publications/testimonies/test-impactkyoto-march25-1999.html). Retrieved on November 15, 2005.

Boffey, Philip. "Talking Points: The Evidence for Global Warming." *New York Times*, July 4, 2006 (http://select. nytimes.com/2006/07/04/opinion/04talking-points.html).

Climate Change 2001: Working Group I: The Scientific Basis, "Estimates of the global methane budget (in $Tg(CH_4)/yr$ from different sources compared with the values adopted for this report (TAR)." (http://www.grida.no/climate/ ipcc_tar/wg1/134.htm#tab42).

IPS. "Independent news on global warming and its consequences." (http://www.ipsnews.net/new_focus/kyoto/index.asp).

IPCC. IPCC Third Assessment Report, 2001. (http://www.grida.no/climate/ipcc_tar/).

Natural Resources Defense Council. "Second Analysis Confirms Greenhouse Gas Reductions in China" (http://www.nrdc.org /media/pressreleases/010615.asp). From the Comprehensive analysis of China's recent economic development and its effects on Chinese emissions.

NOAA. Global Warming FAQ. (http://www.ncdc.noaa.gov/oa/climate/ globalwarming.html #INTRO).

mosnews.com. Russian Government Approves Kyoto Protocol Ratification. September 30, 2004 (http://www.mosnews.com/money/2004/09/30/Kyoto approved.shtml).

UNFCC. (the full text of the convention) (http://unfccc.int/essential_background/convention/background/items/1353.php). Retrieved on November 5, 2006.

UNFCC. Article I (http://unfccc.int/essential_background/ convention/background /items/2536.php).

UNFCC. "An Introduction to the Kyoto Protocol Compliance Mechanism" (http://unfccc.int/kyoto_mechanisms/compliance/introduction/items/3024.php). Retrieved on October 30, 2006.

UNFCC. "Kyoto Protocol: Status of Ratification," 10 July 2006 (http://unfccc.int/files/essential_background/kyoto_protocol/application/pdf/kpstats.pdf). Retrieved on October 30, 2006.

White House Press Release. "President Bush Discusses Global Climate Change." (http://www.whitehouse.gov/news/releases/2001/076/20010611-2.html). Retrieved on November 5, 2006.

Wikipedia. "Global warming" (http://en.wikipedia.org/ wiki/Global _warming).

Wikipedia. "Kyoto Protocol" (http://en.wikipedia.org/wiki/Kyoto_Protocol).

Chapter 18
Assessment of International Trends

From an international perspective, firms are increasingly finding that better management practices can play a key role in addressing many corporate environment problems and pressures that are arising around the globe. The pressures can be both external and internal:

- External pressures are from customers, socially concerned investors, environmental interest groups, and regulators.
- Internal pressures are from firms' own stakeholders, who increasingly expect "their" company to be more "green."

OECD/EIRIS Study Results

A commitment to environmental performance reporting is a strong response to these pressures, but in practice, it is often one of the last elements to be put into place.

Organization for Economic and Cooperative Development (OECD)/Ethical Investment Research Service (EIRIS) conducted a detailed study on environmental management reporting trends. Their data set lays out three major economic regions of the world. Its focus is particularly on large internationally oriented firms, and as such it is not representative of the respective national business communities, but does provide some good soundings on international environmental management trends. The database contains information on environment management systems made publicly available by company annual reports, environment reports, websites, and other materials.

An increasing share of companies in the industrialized economies publishes environmental policy statements (EPS). By September 2003, 58% of all companies in the EIRIS sample had issued statements that meet certain "minimum requirements" (per EIRIS' definitions). The statements typically include:

- A commitment to public reporting
- A commitment to monitoring or audits
- A commitment to use targets
- A reference to allocation of managerial responsibility or a reference to all EIRIS' "key issues"

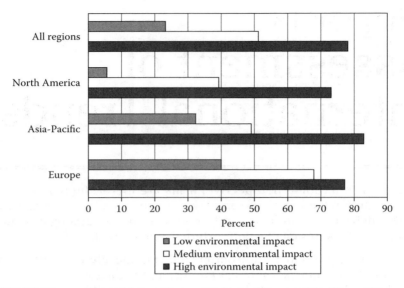

Exhibit 71. Share of enterprises that publish environmental policy statements.

The latter includes such items as suppliers, contractors, resources and materials, energy use and efficiency, emissions to water, emissions to air, transport, waste minimization/reduction/disposal and recycling, packaging, product or stewardship/design, social impact, noise, neighborly concerns, visual blight, employee training, green housekeeping, sustainability, and industry-specific issues.

General results for the three main regions in the index are as follows: 69% of the European companies in the sample have published statements, compared with 62% in the Asia-Pacific region and 44% in North America. The latter dismal number should provide some food for thought with regard to NAFTA's implementation of an environmental management culture. If Mexico were included in the sampling, the results might be even more dismal for North America. However, a word of caution: These observations must be assessed with an understanding of considerable differences within sectors of drivers for issuance of corporate EPSs. Recognize that the publishing of statements involves cost as well as management time for design, publication, and dissemination. Firms for which the environment is not a major strategic or risk management issue are less likely to assume such expenses. In contrast, firms that operate in the more highly regulated environmental impact sectors are more likely to absorb such expenses than firms that operate in a more lax regulatory environment. These national differences in legal environments may make firms more-or-less willing to meet volunteer standards of behavior dependent upon the perception of how key third parties (e.g., the investment community) may assess their actions. Examples of high, medium, and low impact (per the OECD/EIRIS) are as follows:

- *High Impact*—agriculture; air transport; airports; building materials (includes quarrying); chemicals and pharmaceuticals; construction; fast food chains; food, beverages, and tobacco; forestry and paper; major systems engineering; mining and metals; oil and gas; pest control; power generation; road distribution and shipping; supermarkets; vehicle manufacture; waste; and water.
- *Medium Impact*—DIY and building supplies; electronic and electrical equipment; energy and fuel distribution; engineering and machinery; financials not elsewhere classified; hotels, catering, and facilities management; manufacturers not elsewhere classified; ports; printing and newspaper publishing; property developers; public transport; retailers not elsewhere classified; and vehicle hire.
- *Low Impact*—information technology; media; leisure not elsewhere classified (gyms and gaming); consumer/mortgage finance; property investors; research and development; support services; telecoms; and wholesale distribution.

On a more positive note, when looking solely at the high environmental impact sectors there appear to be very few geographical differences. Overall, 78% of all companies in these sectors publish policy statements, and none of the three major regions are far from this average. In contrast, medium impact companies in Europe have much higher rates in this area than companies in the Asia-Pacific or North America. The contrast becomes much more stark when one looks at low environmental impact sectors. At the low sector levels, 40% of the European companies and 30% of Asian companies have published statements versus only 6% of the North American firms in these sectors.

An important question is the content of that EPS. The EIRIS database identified the main elements of an EPS as follows:

- Whether the policy statement makes specific reference to a commitment to comply with the law;
- A commitment to exceed legal requirements; or
- A commitment to best practices.

Exhibit 72 lays out the differences for the three sectors and key nations. Exhibit 73 indicates that EPSs almost unanimously include a commitment to comply with the law (95% of all companies) with very limited geographic differences in this respect. Whereas that might seem to be an obvious corporate policy, it should be recognized that by making such a commitment corporate offices are to a degree emphasizing their personal commitment. When it comes to a "best practice" commitment, a difference

	High Environmental Impact	Medium Environmental Impact	Low Environmental Impact	Total
Europe	**224**	**196**	**78**	**498**
Austria	13	7	2	22
Belgium	8	6	3	17
Denmark	8	10	2	20
Finland	4	2	2	8
France	22	16	7	45
Germany	21	12	3	36
Greece	31	22	15	68
Italy	9	23	5	37
Ireland	3	6	0	9
Luxembourg	0	0	1	1
Netherlands	8	8	2	18
Norway	9	7	4	20
Portugal	5	3	2	10
Spain	12	6	3	21
Sweden	9	13	6	28
Switzerland	11	7	2	20
UK	51	48	19	118
Asia-Pacific	**223**	**217**	**53**	**493**
Australia	31	22	8	61
Japan	156	132	32	320
New Zealand	8	11	3	22
Hong Kong (China)	17	30	5	52
Singapore	11	22	5	38
North America	**183**	**231**	**104**	**518**
Canada	39	32	10	81
United States	144	199	94	437
Total	**630**	**644**	**235**	**1,509**

Exhibit 72. Companies in FTSE All-World Developed Index, by nationality and sector.

develops with the Asian-Pacific region (19%) being higher and Europe (12%) and North America (4%) again trailing badly.

Relative to the third category, "companies' commitments to operate on higher standards than legally required," the Asia-Pacific region showed 74% of companies aiming to exceed legal requirements of their policy statements, compared with 47% in North America and 37% in Europe. However, the measure should be evaluated with caution. Companies located in countries with high legal requirements have less incentive to volunteer only to exceed legal standards, whereas internationally active companies from countries with relatively low legal standards will find it easier, and in some cases may feel under a certain pressure from the investment community, to operate above requirements. Ironically, national and sectoral data

	Nature of Commitment		
	Comply with Relevant Laws	**Exceed Legal Requirements**	**Adhere to Best Practices**
Europe	92	37	12
of which:			
France	86	34	11
Germany	91	45	27
United Kingdom	100	43	6
Asia-Pacific	97	74	19
of which:			
Japan	98	83	21
Australia	98	38	5
North America	94	47	4
of which:			
United States	94	52	2
Canada	93	28	13
Total	95	53	12

Source: OECD/EIRIS

Exhibit 73. Contents of environmental policy statements, all sectors.

show that the "levels of ambition" of policy statements varied little across environmental impact sectors.

An important indicator of the level of corporate commitment is the number of firms that allocate the responsibility for their EPSs to the board level, and as such, strengthening the perception of a high level of managerial interest. The study shows that 89% of the firms have policy statements do so. The Asia-Pacific region showed 95% of the firms allocating responsibility to their boards, whereas a comparatively lower, but still significant, share of 83% tailored this policy.

Another question to consider is whether the EPSs cover the entire business group. This is very pertinent where multinational conglomerates are concerned. The issue of corporate responsibility can be a source of controversy if it is applied differently according to the nationality of operations. The EIRIS database (2003) showed that where a firm has issued a policy statement, as a general rule the whole business group was covered. For nearly 90% of the companies with policy statements in North America, almost all companies (98%) extended their policy statements to their entire multinational groups. In Europe and in the Asia-Pacific region, only 88% and 78% of the surveyed firms did this, respectively.

The EIRIS database showed that a growing minority number of companies are also signing up for voluntary environmental initiatives. Exhibit 74 provides survey results regarding the participation of companies in either of four such voluntary initiatives, namely the International Chamber of Commerce's (ICC) Business Charter for Sustainable Development, the

	Impact Sector			
	High Environmental Impact	Medium Environmental Impact	Low Environmental Impact	Total
Europe	39	33	5	31
of which:				
France	32	44	14	33
Germany	86	67	33	75
United Kingdom	43	27	0	30
Asia-Pacific	23	11	13	17
of which:				
Japan	31	15	16	23
Australia	10	9	25	11
North America	24	9	0	13
of which:				
United States	26	8	0	12
Canada	15	19	0	15
Total	29	17	5	20

Includes UNEP F1, responsible care, ICC, and ceres.

Exhibit 74. Signatories to Voluntary Initiatives.

Coalition for Environmentally Responsible Economies' (CERES) Corporate Environmental Reporting Requirements, UNEP's Finance Initiative, and the chemical industry's Responsible Care.

A similar trend was reportedly seen relative to the issuance of EPSs. It should also be noted that in both cases, there was a significant drop in participation for all regions (issuance) as one assessed high versus medium versus low impact companies. It may be indicative that the lower impact companies see little need to make announcements over and beyond a normal policy statement. Of particular note are the dismal numbers for North American companies versus their European counterparts.

Still, it should be recognized that the publication of an EPS is only a part of the process to assess company environmental performance and management. The pressure of formal management control practices such as an environmental management system (EMS) is also a strong indication of environmental performance and commitment. Though the EMSs may vary widely in details from organization to organization, most typically include the following parts: an EPS; an initial review; environmental objectives and targets, implementation procedures, internal monitoring and auditing; and internal reporting.

OECD/EIRIS also found that the implementation of environmental management systems followed largely the same sectoral and national patterns as the issuance of EPSs. European firms were more likely to implement

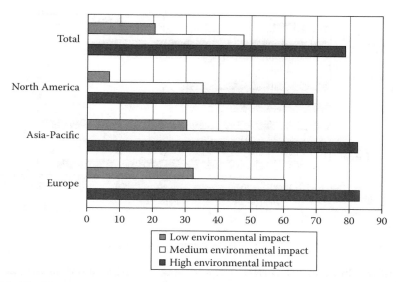

Exhibit 75. Share of enterprises that have implemented environmental policy statements.

management systems than North American and the Asia-Pacific region counterparts (66% for European companies, compared to 62% for Asian-Pacific and 41% for North American companies). The United Kingdom and Germany (both with 86%), followed by France (82%), led the pack while the lowest rates of implementation from the sample were found in Singapore (18%), Hong Kong (19%), and Greece (23%). However, similar to the situation with EPSs these differences diminished when looking only at high environmental impact sector companies (Exhibit 75) in Europe and the Asia-Pacific (83%) and North America (69%).

Some firms choose to set up self-designed EMSs. By doing so, companies can tailor the individual company requirements and problems. Other companies adapt standardized environment management standards. The advantages of tailor-made management systems on the one hand and standardized systems on the other have been discussed in relation with other areas of management, and there appear to be similar discrepancies with regard to EMSs. Going the standards route enhances the credibility of a firm's environmental measures, assuming the selected management standards are widely accepted. Standardized systems also provide quick and inexpensive access to advanced management techniques. However, a disadvantage of standardized systems is that they may not be entirely suited to individual company needs.

The most common standardized EMS, ISO 14001, is an international environmental management standard. ISO 14001 was first published in

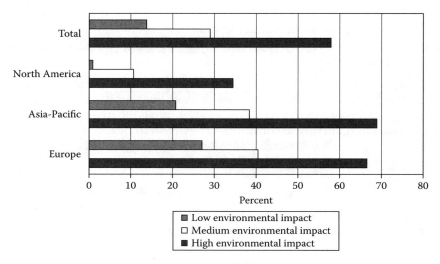

Exhibit 76. Share of enterprises that undertake environmental performance reports.

1996. Later, other more encompassing environmental standards incorporated ISO 14001 as a key element. An example is the Eco-Management and Audit Scheme (EMAS), the E.U.-supported management system and certification scheme. EMAS goes beyond the scope of ISO 14001 by establishing minimum standards for auditing and environmental reports.

The OECD/EIRIS study found that two-thirds of the companies that have EMSs in place have either an ISO 14001 certification of their system or have implemented the ISO standard as part of an EMAS certification (Exhibit 76); the other third have EMSs tailored to the individual enterprises. By and large, North American firms have the highest share of tailored EMSs (50%). Also, a significant percentage of the tailored North American EMSs is not deemed to be compatible with ISO 14001. The EIRIS database showed that almost 70% of firms with EMSs in place in 2003 undertook environmental auditing. The tendency was lower in European countries (58%) and higher in the Asia-Pacific region (84%). In particular, the Japanese business sector had a high share of activity.

Companies that have high standards for environmental management typically guard themselves against being victimized by shortfalls in the environmental performance of their suppliers and contractors. Supply chain auditing has emerged as a powerful tool for providing corporate buyers with comprehensive environmental information on the products, components, or materials they produce, and in so doing, protecting the purchasing company. Supply chain auditing also provides an impetus for change among small and medium-sized suppliers that, on their own initiative, might not be as proactive. However, at this juncture per OECD/EIRIS only a limited number (14%) of the companies that have EMSs engage in environmental supply chain audits, with just 31% of the North American enterprises and 38% in the case of the United States.

Ironically, North America leads on this initiative. Just as ironic, the study found that high-risk impact industries are less likely to implement supplier audits. Apparently, low-risk companies are more concerned about the risk of being made "guilty by association." They tend to devote their resources instead to their own intra-firm environmental performance.

The OECD/EIRIS study found that in economies where environmental management practices have been widespread, the demand for high-quality environmental reports was increasing with companies facing increasing pressure to publish a thorough report on their environmental performance. The latter includes quantitative information going back several years and references to negative experiences.

However, in the absence of internationally agreed-upon reporting standards the content of reports ranges from general information to full-scale sustainable development reporting. Studies in 2001 and 2003 undertaken by OECD concluded that the area of environmental performance reporting is "the least common of the three environmental practices considered" (e.g., EPS, EMS, and environmental performance reports). However, the studies found that approximately two-thirds of the companies operating in high-impact sectors in both Europe and the Asia-Pacific undertake environmental performance reporting; in North America, only one third of the firms in the high-risk industry sector do so. The countries in which performance reporting is the most pervasive are Germany (86%), the United Kingdom (71%), and Japan (69%).

Widely accepted standards to help firms decide what information should be included in their environmental performance reports (EPR) are lacking. The Global Reporting Initiative (GRI) is a multi-stakeholder initiative set up by CERES. GRI's ultimate aim is to bring environmental performance reporting to the same level as financial reporting by developing a set of guidelines for companies to follow. Other guidelines, frameworks, and standards include Social and Ethical Accounting, Auditing and Reporting, Corporate Community Investment by the London Benchmarking Group, Fondazione Eni Enrico Mattei, Health, Safety, and Environmental Reporting Guidelines by the European Chemical Industry Council, and the Public Reporting Initiative.

Given the absence of an agreed-upon standard for environmental reporting, firms are left to their own initiative with regard to the scope and depth of their reporting. Exhibit 77 lays out four indicators of differences between the contents and scope of existing EPRs. Clearly, there is a growing need for

	Publish Quantitative Data	Compare Performance with Targets	Rely on Third-Party Verification	Environmental Cost Accounting
Europe	85	56	46	43
of which:				
France	92	44	36	32
Germany	94	71	29	48
United Kingdom	89	73	44	35
Asia-Pacific	97	86	29	30
of which:				
Japan	100	90	29	29
Australia	85	85	46	85
North America	94	45	11	21
of which:				
United States	93	36	10	10
Canada	100	67	14	57
Total	91	67	34	34

Exhibit 77. Nature of companies' environmental performance reports (percentage share of companies that issue EPR/EPS).

EPRs to be developed that have quantitative information publicly available and comparable. High-risk environmental impact industries in almost all countries are more likely to publish quantitative information.

EPRs should include information that allows stakeholders to monitor their progress toward implementing higher environmental standards. For example, performance targets could be presented along with quantitative information to allow a comparison of actual data with past targets.

Third-party independent verification of the environmental management system is a critical way to establish credibility regarding environmental performance. Verification also provides management with a level of confidence that the reporting system is accurate. Verification should be conducted by qualified external parties that are also independent from the data collection and report production process. In the OECID/EMIS study, only about one third of the companies issuing performance reports relied on third-party verification of those reports' contents.

Why has third-party verification not gained wider acceptance? The lack of an internationally recognized assurance standard may be a reason. Whereas the AA1000 Assurance Standard was launched in April 2003 by AccountAbility, the success of the initiative is still an open question. In summary, some of the OECD/EIRIS key observations about corporate environmental management that can be derived from the survey results follow.

EPSs were published by almost 60% of the surveyed companies, with the highest percentage being found in Europe (69%). Within the high-risk impact industries, almost 80% of all firms surveyed published policy statements with little variation across major regions. Almost all policy statements of the surveyed companies included a commitment to comply with the law. The signup to voluntary environmental initiatives follows a trend similar to that of the issuance of policy statements. One third of the surveyed companies were signatories to voluntary environmental initiatives. Most of the signatory companies operated in the high environmental impact sectors. Also, the acceptance of such initiatives is somewhat higher in Europe than elsewhere.

The implementation of EMSs shows trends similar to the issuance of policy statements. European enterprises are more likely to do so (66%), with North America lagging (41%). Also, EMSs are far more prevalent among companies operating in the high-impact environment sectors. ISO 14001 appears to be the prevailing EMS across the globe. Two-thirds of all EMSs are either ISO certified or follow a standard that encompasses ISO 14001. The EMSs that are not ISO certified—and according to EIRIS' assessment, not ISO compatible—are not prevalent in North America. Over 70% of companies with EMSs engage in environmental auditing, and supply chain audits are particularly popular among North American companies.

Environmental performance reporting is less common than the other practices considered in the OECD/EIRIS report. However, it should be recognized that environmental reporting is relatively new compared to policies and management systems, so it is not necessarily surprising that there is a lag.

In the absence of an internationally agreed-upon standard, the scope of environmental reporting varies greatly. Most of the reporting companies (over 90%) publish quantitative data that allow comparisons of performance intra-industry and over time, and two-thirds of performance reports include some comparison of companies' environmental performance relative to previous targets. However, only one third of the EPRs are subject to third-party verification of their content.

Occupational health and safety systems have historically been less standardized than environmental systems. However, the ISO-compatible standard OHSAS 18001 appears to be gaining widespread acceptance. Studies (e.g., EIRIS) indicate that only 20% of the companies in the sample display "clear evidence" of having an occupational health and safety system in place. The geographic differences are significant. A total of 34% of European enterprises fall into this category, compared with 15% in North America and 9% in the Asia-Pacific region.

The principle underlying the OECD/EIRIS classification system is that a sector's overall environmental impacts should be assessed in relation to its size. For each sector, direct impacts relating to climate change, air pollution, water pollution, waste, and water consumption were reviewed. Impacts arising indirectly through upstream (supply chain) or downstream (product

	Clear Evidence	Some Evidence	Little or No Evidence
Europe	34	31	35
of which:			
France	49	33	18
Germany	22	46	32
United Kingdom	56	32	11
Asia-Pacific	9	23	67
of which:			
Japan	5	26	69
Australia	30	21	49
North America	15	25	60
of which:			
United States	16	25	60
Canada	12	27	60
Total	20	26	54

Exhibit 78. Evidence of the presence of an occupational health and safety system (percentage share, by country or region).

life cycle) were also considered, mainly in qualitative terms. The basic indicator used was a ratio of environmental damage to economic significance.

Survey on the State of Global Environmental and Social Reporting

Another survey of interest, "The State of Global Environmental and Social Reporting: The 2001 Benchmark Survey," was conducted to help companies gauge their progress on reporting and identify areas that need improvement or showed strong gains in environmental reporting among the larger multinational segments. Conducted by CSR Network Limited, a U.K.-based international consultancy, the survey examined the 100 largest firms listed in *Fortune Magazine*'s August 2000 Global 500.

The most notable result of the survey was that for the first time, half of the largest 100 (G100) corporations produced global environmental reporting. This figure is up from 44% in 1999. The survey showed that reporting on global EMS specifications and standards have remained high after an upsurge in 1999. Approximately 44% of companies disclosed EMS specifications and 45% disclosed global EMS standards. Among these companies, 34% are reporting EMS-related goals as well.

Supply chain environmental management (SCEM) reporting also increased since 1999. Nearly a quarter of the G100 disclosed global data on greenhouse gas emissions, and 14% addressed the environmental impacts, distribution, and transportation. Toyota, Mitsubishi Electric, and Amoco were cited as examples of companies that had the best reporting of greenhouse gas emissions. The study revealed that over half (53%) of the companies reported on their social programs policies as well. However, the study noted a divergence in reporting methods.

Emerging China and India Environmental Issues

China and India are clearly emerging as the critical players in the Asian environmental scene. Given their current exempt status from Kyoto, they are a huge hole in the environmental control for Asia in general. For decades, industry has been the main source of pollution in China. However, several striking events of the last year (e.g., major spills in waterways) have spurred a "green trend" in China. There appears to be a new determination on the government's part to make changes. To that end, mainland Chinese authorities have decided to promote mechanisms that incorporate environmental concerns into the internal management of enterprises. This is manifested in the rapid adoption of the ISO 14000 standards. Reported analysis of the contents of the EPSs shows that conformance to the relevant requirements of both the mandatory ISO 14001 standard and the non-mandatory ISO 14004 standard is far from impressive, and that the facilities in our sample seldom went beyond the minimum requirements. Still, there is an emerging need for multinationals doing business in China to engage in aggressive environmental vendor surveys of Chinese facilities to aid in this turnaround.

As for India, they too are slowly addressing environmental concerns. Of particular significance is the "Asian Brown Cloud" phenomenon that enters on the India subcontinent. However, it is not apparent that the Indian government has, as of yet, taken any aggressive steps to address their environmental energy crisis.

The "Asian Brown Cloud" is a dense blanket of pollution hovering over South Asia. Two hundred scientists have warned that the cloud—estimated to be two miles thick—is responsible for hundreds of thousands of deaths a year from respiratory disease. By slashing the sunlight that reaches the ground by 10% to 15%, the choking smog has also altered the region's climate, cooling the ground while heating the atmosphere. The potent haze lying over the entire Indian subcontinent is a virtual cocktail of aerosols, ash, soot, and other particles. The haze extends far beyond the study zone of the Indian subcontinent, toward East and Southeast Asia. Not only has it been discovered that the smog cuts sunlight and heats the atmosphere but it also creates acid rain, a serious threat to crops and trees, and contaminates oceans.

What are the true costs of the environmental externalities of emerging economies such as China and India? There is increasing consensus that when you subtract the costs of air and water pollution and related health impacts, many of the glowing growth rates of theses economies may not be showing growth at all. Recognize that the polluters at present bear the costs of these environmental externalities. Instead these costs are borne by the public through increased illness and death rates. It is estimated that in China alone, 400,000 Chinese die of air pollution related illnesses each year and almost 300 million Chinese do not have access to clean water supplies. As these numbers mount, the environmental externality piper will eventually need to be paid by the polluters and the "environmental externality gap" for these nations will be extremely expensive to close and ultimately, may wipe out the current labor cost advantages that these nations enjoy in the world market. Recognize too that world events, such as China's hosting of the Olympics, can be driver into forcing environmental progress.

Kyoto Protocol Debates

The Kyoto Protocol is an international treaty by many different countries that would regulate the amount of industrial gases that one country could emit. Whereas the current Bush administration conceded in generic fashion that the global warming crisis is a very important one, the administration continues to focus on how (in their view) scientists may agree that human activity has contributed to the increase in climate change but are not sure about how much of this activity is natural. In addition, the Bush administration defends their reluctant (if not adversarial) stance to not be a part of the Kyoto Protocol because it is "fatally flawed in fundamental ways." Whereas it is generally conceded that the United States is the number one emitter of greenhouse gases released in the world, the administration points out that other industrial countries are exempted from the protocol, including China, Germany, and India. To put it simply, the administration takes the view that the rest of the world emits 80% of all gases in the world and the global warming crisis needs a 100% effort. The administration also believes that the target proposed for the United States would potentially be too strenuous on the American economy.

How critical is global warming? Rajendra Pachauri, the chairman of the Intergovernmental Panel on Climate Change (IPCC) offered that the global environment has "already reached the level of dangerous concentrations of carbon dioxide in the atmosphere," and called for immediate and "very deep" cuts in the pollution if humanity is to "survive" (Geoffrey Lean). His views should be given added weight, considering the Bush administration supported his appointment to chairman based on his previously perceived less-aggressive beliefs on climate change. The former chairman of the IPCC before Pachauri was Robert Watson, the chief scientist of the World Bank. Watson was replaced at the formal request of the U.S. because of his more aggressive stance (Lean). It should be noted that this came right after Exxon, a major American oil company that is opposed to immediate action on global warming, voiced the same complaint about Watson. However, Pachauri has veered from a less-aggressive stance because of the mounting evidence about the effects of global warming. (IPCC is considered to be the world's foremost authority on climate change.)

According to Pachauri, the polar ice caps are 40% thinner today than they were in the 1970s, and the ice caps may disappear by the year 2070.

Also, coral reefs have been dying off at an alarming rate because of the increase in water temperature. Up to a quarter of all coral reefs in the world have been destroyed. It is ironic that the Bush administration finds itself at odds with Pachauri's evolving stance as they clearly aided his appointment in 2001. The final IPCC report issued in early 2007 was a 1,500 page scientific assessment based on 29,000 sets of data collected in the last five years. The report is the most comprehensive scientific review on the impact of global warming due primarily to man-induced carbon dioxide emissions. The report concludes that by 2020 as many as 250 million people across the globe will likely be exposed to water shortages and severe cuts in food production, with Africa being the most severely hit area. The report adds that the world will face greater threats of flooding, severe storms, and erosion of coastlines as well. The report's 21-page summary was a clear statement prepared as a policy guide for governments and reflected the input of representatives from 20 nations that attended the IPCC meeting.

State of the U.S. Environmental Policies

International studies could lead to the question: "What about the U.S.?"

Heather Taylor, Deputy Legislative Director of the Natural Resources Defense Council (NRDC), prepared a detailed report laying out the lack of consideration the Bush administration has given to environmental causes. According to the NRDC, Bush's new spending plan for 2007 will cut overall environmental funding by 13%. On average, most federal agencies received a half-percent reduction in their budget, whereas the EPA was cut 4%. One of the areas of the environment that has been hit hardest by Bush's budget cuts is the Clean Water Revolving Fund. Bush plans to cut $200 million from last year's budget and has cut $1.4 billion from funding for clean water since 2002. Bush continues to cut the budget even though the EPA believes they will need $19 billion dollars annually for the next 20 years to solve the U.S.'s clean water problem.

From a clean air perspective, there also is reason for concern within the U.S. It is estimated that 150 million Americans live in areas where the air is considered to be unhealthy. While the Bush administration claims that funding has increased to reduce diesel emissions into the air, it has done so by cutting funding from other areas that fund local clean air and water initiatives. Sadly, the U.S. has taken a back seat in environmental affairs during the Bush administration. Basically, the Bush administration is taking money away at the local level, even though these local agencies need the money to improve air quality under the Clean Air Act.

"Greenwash" versus "Green Machine" Debate

There has been increasing discussion regarding the true commitment behind the emerging environmental glossy reporting structures. To some extent, this can be summarized as a "greenwash versus green machine" debate, and corporations are forewarned that glossy claims may be subject to scrutiny and dispute.

For example, in an article entitled, "A convenient confusion" in the June 2001 *New Internationalist*, Kenny Bruno "examines what's behind the glossy, eco-friendly ads the oil companies are churning out... There's even more confusion about the solution. In part this is because it's hard to face the realization that oil and gas, the lifeblood of industrialized economies, are also the main source of carbon dioxide, the major greenhouse gas." A couple of examples of the debate are the experiences of ExxonMobil and Wal-Mart.

ExxonMobil

Initially, ExxonMobil maintained that fears of global warming "are based on completely unproven climate models or more often on sheer speculation without a reliable scientific basis." Their initial position was to fund studies that challenge global warming. However, in recent months they have retracted from their initial adversarial position. *Mother Jones Magazine* traced nearly $9 million of ExxonMobil's corporate contributions made between 2000 and 2003 to "more than 40 think tanks; media outlets; and consumer, religious, and even civil rights groups that preach skepticism about the oncoming climate catastrophe." However, in recent months they have retracted from this hard position. Possibly, Exxon's shift may be the result of a recommendation on July 12, 2006, by a coalition of major environmental and progressive groups representing more than six million members, that called for a consumer, investor, and employee boycott of ExxonMobil. Whatever the reason for the shift, ExxonMobil appears to be making a strategic shift in how they are confronting the issue, and this is a prime example of the emerging critical importance of strategic corporate environmental policy and positioning.

Wal-Mart

Another recent example of a significant shift in strategic environmental policy is Wal-Mart. In November 2005, Wal-Mart set several ambitious goals:

- Increase the efficiency of its vehicle fleet by 25% over the next three years, and double its efficiency in ten years
- Eliminate 30% of the energy used in stores
- Reduce solid waste from U.S. stores by 25% in three years

Wal-Mart also committed to investing $500 million in sustainability projects. The company is the biggest private user of electricity in the U.S.; each of its 2074 "supercenters" uses an average of 1.5 million kilowatts annually. Simple aggressive energy efficiency would provide both an enormous environmental plus cost benefit. In contrast, Wal-Mart has been credited recently with a "green machine" initiative that is increasingly winning over skeptics ("The Green Machine," by Marc Gunther, published in the July 31, 2006, *Fortune Magazine*). And because of the extraordinary clout Wal-Mart wields with its 60,000 suppliers, it could make even a greater difference by influencing their practices.

Prior to the recent turn of events, the company's environmental record had no reason to boast. It had paid millions of dollars to state and federal regulators for violating air- and water-pollution laws. Wal-Mart found that up to 8% of shoppers had stopped patronizing the chain because of its reputation. Then Wal-Mart hired BluSkye, CI, and Ellison's management consulting firm, to measure Wal-Mart's environmental impact. About a dozen people from BluSkye, CI, and Wal-Mart spent nearly a year measuring the company's environmental impact and pretty quickly discovered waste that Wal-Mart's own cost-cutters had overlooked. Today, Wal-Mart has 14 focused networks: Facilities, Internal Operations, Logistics, Alternative Fuels, Packaging, Chemicals, Food and Agriculture, Electronics, Textiles, Forest Products, Jewelry, Seafood, Climate Change, and China. Wal-Mart began to add organic items such as food items and organic cotton. LED lighting was put into Wal-Mart's buildings.

References

"'Asian Brown Cloud' poses global threat," posted August 12, 2002, on CNN.com/WORLD. (http://archives.cnn.com/2002/WORLD/asiapcf/south/08/12/ asia.haze/). Retrieved on January 8, 2007.

China Environmental Protection Foundation. "Introduction to the China Green Territory Project," China Green Map (http://www.cepf.org.cn/en_ leading/green.asp). Retrieved on January 8, 2007.

Gunther, Marc. "The Green Machine." *Fortune Magazine*, July 31, 2006 (http:// money.cnn.com/ magazines/fortune/fortune_archive/2006/08/07/8382593/index.htm). Retrieved on January 8, 2007.

Hart, Stuart, "Beyond Greening: Goal and Strategies for a Sustainable World," Harvard Business Review, Jan.-Feb. 67–76, 1997.

Jubak, Jim, "How Long Can China Pollute for Free?" Jubak's JournalæMSNBC, February 9, 2007.

Lean, Geoffrey, *Independent News & Media* (UK), Ltd. 23 January 2005. (London: Independent News & Media).

MSNBC NEWS Service, "Expert Issue New Climate Warming: Brussels, Belgium," April 6, 2007.

OECD Secretariat and EIRIS. "An Overview of Corporate Environmental Management Practices, Joint Study by the OECD Secretariat and EIRIS."

Taylor, Heather. NRDC Media Ctr. Press Statement. 8 Feb. 2006, Washington, D.C.

The Green Life. "The Lie of the Tiger," posted July 2005 (http://www.the greenlife.org/ greenwasherjuly2005.htm). Retrieved on January 8, 2007.

Chapter 19
Summary

As shown in Exhibit 79, the three-point environmental management assessment provides the company with three different perspectives for environmental management performance. Thus, corporate management receives feedback that reflects not only the performance from within the organizational standpoint, but also takes into account the organization's position relative to its business community as a whole and reflects the expectation and the realities of public stakeholders.

But EHS performance is not strictly an EHS management mission; it is a company-wide mission and should be recognized as such. To that end, annual compensation adjustments for all employees should be tied in part to environmental performance (i.e., no violations). An EHS awards program ensures that both EHS and non-EHS employees are recognized for suggestions, innovation, and superior performance.

Also recognize that it is critical that an environmental management assessment be conducted in a truly independent fashion, from day-to-day corporate to line-responsible personnel. To that end, an environmental management assessment should always be conducted for and reported to the appropriate subcommittee of the board of directors. However, there may be circumstances when a board may wish to have the findings presented through the auspices of the company's outside legal counsel for confidentiality purposes. Still, key findings should always be communicated to appropriate, responsible environmental and operational management parties.

But recognize that the goal is not just environmental management; it is the degree to which sound environmental management contributes to sustainable development, as defined by the United Nations World Commission on Environment and Development. This calls for "development that meets the needs of the present without compromising the ability of future generations to meet their own needs." Sustainability can also be defined as the ability of the company to continue into the long-term through excellence in performance and stewardship, as depicted in Exhibit 80.

To look at it another way, the concept of sustainability can be visualized as three intersecting circles—representing financial, environmental, and community performance—that overlap into realms of efficiency and accomplishment.

Exhibit 79. Different perspectives of environmental management performance.

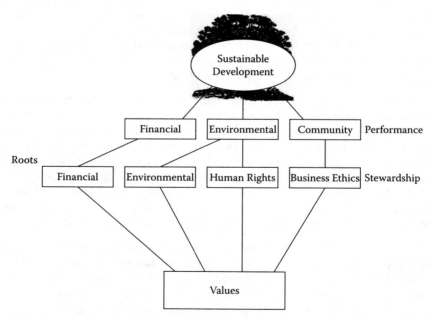

Exhibit 80. Sustainable development.

In summary, a culture must be fostered wherein EHS is recognized as good for sustainable business. Environmental management is a venue that requires not only aggressive access to policy-setters and regulators but concrete efforts to improve relations with external stakeholders. In time, strong environmental management may offer increased business

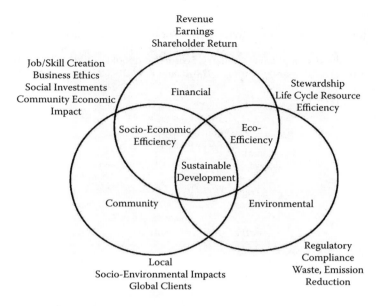

Exhibit 81. Understanding sustainable development.

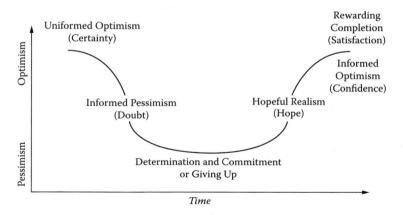

Exhibit 82. Recognize the emotional cycle for environmental management change.

opportunities and enhanced worker morale. Last, corporate management must recognize that effecting a strong environmental management program change calls for an emotional cycle to be worked through, as in any organizational change. Strong corporate leadership is paramount to ensure successfully transitioning the cycle.

References

ISO. "Environmental Management Systems—Specification with Guidance for Use." Reference number ISO 14001:1996(E). (West Conshohocken, PA: ASTM. 1996.)

U.S. EPA's Office of Compliance: Industry Sector Notebooks; EPA/310-R-95-001 through 95-018; EPA/310-R-97-001 through 97-010; and EPA/310-R-98-001 through 95-008, 2001.

Index

A

AA1000 Assurance Standard, 184
Accounting systems, 72–75
Acid rain, 172, 187
Activist groups, 106–107
Activity-based accounting, 73
American Petroleum Institute, 56, 94
Amoco, 56, 186
Appliance recycling, 2
ARCHIE model, 125
Arsenic, 137–138
Asian Brown Cloud, 187
Assessment, *See* Environmental
    management assessment;
    Risk assessment
Asset management, 75–77, 80–81, 101–102
Association of British Insurers, 100
ASTM standards, 73, 151
Atmospheric emissions, 167–174, *See also*
    Greenhouse gas emissions;
    Pollution controls
    Asian Brown Cloud, 187
    Clean Air Act and amendments, 59, 127, 172
    emission reduction program, 63–64
    Kyoto Protocol, 76, 167, 169, 172–173
    startup- and shutdown-related, 129
    up-the-stack emergencies, 127–130
    U.S. environmental policies, 76, 172, 189
Audit program, 65–70, *See also*
    Environmental management
    assessment
    business performance reviews, 69
    control methods, 67
    environmental management review, 70
    root cause analysis, 69–70
    scope of, 66
    site visits, 68–69

B

Balance sheet risk, 84
Banks, *See* Financial institutions
Baselines, 58–59
    audit program, 66
    risk assessment, 117, 123
Benchmarking survey, 13–14, 16, 87–98
    environmental database, 93–94
    key factors, 87–88
    management-oriented approach, 89–91
    technical-based approach, 91–95
Benzene, 127
Bergsoe Metal Corp., 148
Bhopal, India, 35
Bond financing, 80
BS 8800, 135
Business Charter for Sustainable
    Development, 34, 36, 179
Business performance reviews, 69–70
Business Sector Environmental Accounting,
    73, 75
Business sustainability risk, 84

C

California reporting guidelines, 64
Canada, 168, 173
Cancer risk, 123
Capital cost risk, 84
Carbon dioxide (CO_2), 2, 63, 167, *See also*
    Greenhouse gas emissions
    brokers, 169
    global warming and, 188–189, *See also*
    Global climate change
Carbon monoxide, 127
Carbon sequestration, 168, 172
Carpeting recycling, 155
CERCLA, 149, 151
CERES, 34, 35, 180
Certification, *See* Environmental certification
    schemes
Characteristic wastes, 139
Chemical hazard classification, 139
Chemical Manufacturers Association (CMA),
    34
Chemical toxicity values, 119
China, 187–188
Citizens' advisory groups, 44
Clean Air Act, 59, 127, 172
Clean Development Mechanism (CDM), 173
Clean Power Act, 167
Clean Skies initiative, 76, 167
Clean Water Act, 59
Clean Water Revolving Fund, 189
Coalition for Environmentally Responsible
    Economies (CERES), 34, 35, 180

Commitment, *See* Corporate commitment
Communication, 43, *See also* Environmental
    performance reporting
    consistent communications plan, 43
    criteria, 20
    external, 43–45
    hazardous communications, 138
    incident command system, 134
    informal networks, 46
    integrating environmental and financial
        performance, 159–161
    internal, 46–48
    ISO 14001 and, 39
    issue management, 48–49
    sustainable development reports, 45
    websites and, 43–44
Competition, environmental-based, 5, 6
Competitive advantage, *See* Environmental
    management as business asset
Comprehensive Environmental Response
    Compensation and Liability Act
    (CERCLA), 149, 151
Consumer influence, 3, 106
Contingency plans, 126–127
Contractor support, 17, 19
Control systems
    audit program and, 67
    criteria, 21
Conventional commercial loans, 79
Coral reefs, 189
Corporate climate change profile, 169
Corporate commitment, 5, 31, 195
    criteria, 20
    environmental policy, 33–34
    executive level management, 31–33, 39
    Performance Track Corporate Leader, 40
    responsibility for performance reporting,
        179
    strategic environmental planning, 40–42
Corporate environmental culture, 53, 194
Corporate environmental management, 5
    assessment, *See* Environmental
        management assessment
    business advantage, *See* Environmental
        management as business asset
    business costs and benefits, 67–68
    certification, *See* Certification
    consciousness raising, 84
    financial issues, *See* Financial
        institutions; Financial
        management; Investment and
        environmental management

global issues, *See* Global climate change;
    International environmental
    management trends
green wall, 84
headquarters control over, 27
outside the mainstream, 17, 20
"plan, do, check, act" model, 67
proactive environmental management, 84
program vs. project management, 20
risk assessment, *See* Risk assessment
stakeholder influence, 106
Corporate environmental policy, 33–34,
    175–180
Corporate health and safety, *See* Occupational
    health and safety
Corporate operational control, 27
Cost feasibility, 24
Cost management, 72–75
CSR Network Limited, 186
Cultural diversity, 104

D

Databases, 93–94
Decontamination, 144
Denmark, 167, 173
Design and construct phase, 104
Design for the environment (DfE) approach, 5
Documentation, 10
    stiffness factor, 25
Domini Social Index, 157
Dose, 119, 136
Dose-response, 118–119, 136
Dow Jones Sustainability Index, 100
Due diligence, 153

E

Earth Summit, 158
Eco Balance, 74
Eco-funds, 6–7
Ecological risk assessment, 120–122
Eco-Management and Audit Scheme (EMAS),
    182
EcoVALUE'21, 85
Electric Power Research Institute, 95
Emergency response planning and analysis,
    125–134
    allowable excess emissions, 128–129
    ARCHIE model, 125
    contingency plans, 126–127
    effective communication, 134

explosion and vapor dispersion models, 126
incident action plan, 132
incident command system, 130–132
logistics, 132
potential incident management weaknesses, 130
preparedness, 134
regulatory requirements, 126
root cause analysis, 129
tactical incident action planning, 133
up-the-stack emergencies, 127–130
Emission controls, *See* Pollution controls
Emission factor sources, 64
Emissions, atmospheric, *See* Atmospheric emissions; Greenhouse gas emissions
Emission source identification, 63
Emissions reporting, 127–130
Emissions trading, 76, 167–169, 172–173
brokers, 169
internal trading markets, 168–169
Employee health and safety, *See* Occupational health and safety
Employee rewards, 38
Employee-Right-to-Know program, 138
Employee training, 38
End-of-pipe pollution controls, 1–2, 56, 155
Energy conservation, 3
Environ, 182–183
Environmental accounting, 72–75
Environmental activist groups, 23, 106–107
Environmental asset management, 75–77, 80–81, 101–102
Environmental Banking Association (EBA), 36
Environmental-based competition, 5, 6
Environmental capital equipment leasing, 81
Environmental certification schemes, 6, 37, 67–68, 70, 183, *See also* ISO 14001; Third-party verification
competition among, 107
preparedness, 134
Environmental consciousness raising, 84
Environmental cost management, 72–75
Environmental credit assets, 76
Environmental culture, 53, 194
Environmental databases, 93–94
Environmental financial management, *See* Financial management
Environmental health and safety (EHS), *See* Occupational health and safety
Environmental impact assessment, 65, 185–186

of environmental performance-reporting firms, 176–180, 184
global climate change, 171–172, 188–189
green purchasing vs., 3
measurement systems, 38
RICOH approach to environmental accounting, 74
transnational responsibility, 159
Wal-Mart, 191
Environmental issue management, 48–49
Environmental liability for lenders, *See* Lender liability
Environmentally responsible investment, *See* Investment and environmental management
Environmental management as business asset, 32, 72, 77, *See also* Investment and environmental management; Sustainable development
asset management, 77, 101–102
cost savings, 159
integrating environmental and financial performance, 157–164
opportunity vs. liability, 83, 156
Environmental management assessment, 5, 9–11, 13, 193, *See also* Risk assessment
analytical framework, 19–21
audit program, 65–70
benchmarking, 13–14, 16, 87–98, *See also* Benchmarking survey
closure phase, 105
contractor support, 17, 19
corporate level and, 10
design and construct phase, 104
documentation, 10
environmental impact assessment, 65, 185–186
environmental management review, 70
evaluation phase, 104
external survey, 13, 15–16, 87, 99–108, *See also* External scan
implementation phase, 104
independent technology survey, 15–16, 107
internal survey, 13–15, 27–29, 87
natural resource damage assessment, 15–16, 109–115
operations, 29, 104
project life cycle review, 103–105
regulatory model, 25
site visits, 15, 68–69
stages of excellence model, 17–19
strategic management, 101

uncertainties and, 10, 24
Environmental management information
    systems, 70–72
Environmental management-oriented survey
    approach, 89–91
Environmental management system,
        See Corporate environmental
        management
Environmental operational systems, 53–59
Environmental performance indicators, 37
Environmental performance reporting, 6, 44,
    155–156
    business performance reviews, 69
    corporate commitment indicators, 176
    Global Reporting Initiative, 183
    "greenwash" vs. "green machine" debate,
        189–191
    guidelines, frameworks, and standards, 183
    high environmental impact sectors,
        176–178, 184–185
    integrating environmental and financial
        performance, 151–161
    international trends, 183–185
    supply chain environmental management
        reporting, 186
    10-K reporting, 72
    voluntary environmental initiatives,
        179–180
Environmental policies
    corporate, 33–34, 175–180
    international trends, *See* International
        environmental management
        trends
    U.S., *See* U.S. environmental policies
Environmental program implementation, 21,
    51, 104
    emission reduction program, 63–64
    internal integration, 52–53
    international trends, 180–183, 185
    operational systems, 53–59
    organization and staffing, 51–52
    pollution prevention opportunities, 55–59
    risk assessment systems, 60–61
    waste minimization programs, 61–63
Environmental rationality, 5
Environmental restoration sites, 110, 112
Environmental risk assessment, *See* Risk
    assessment
Environmental risk management, 22–25,
    60–61, 117
    at banking institutions, 147–164, *See also*
        Financial institutions
    financial risks, 83–84, *See also* Financial
        management; Lender liability

identifying management alternatives, 24
    operations-level responsibility, 29
    oversight committee, 32–33
    perceived risk, 22, 23
    prioritizing risks, 23
    regulatory risk, 22, 23
    risk types, 60
    technical risk, 22
Environmental site assessments, 68–69, 151
Environmental stewardship, 104
Environmental value fund approach, 162
Equipment leasing, 81
Equity banking, 154–157
Ethical Investment Research Service
        (EIRIS)/OECD study, 175–186
Evaluation phase, 104
Exclusion zone (EZ), 144
Executive management commitment, 31–33,
    39
Exposure assessment, 118–119, 121
Exposure-point concentration, 119
Exposure routes, 136
External communication, 43–45
External scan, 13, 15–16, 87, 99–108
    challenging business environment, 102–103
    global impacts, 99
    independent technology survey, 15–16, 107
    phases, 104–105
    project life cycle analysis, 103–105
    sustainable development, 99–103
ExxonMobil, 190
Eye protection, 140

F

Feasibility, 24
Federal Deposit Insurance Corporation (FDIC),
    150
Federal Reserve, 150
Fiduciary responsibility, 163–164
Financial institutions, 6, 147–164, *See also*
        Investment and environmental
        management
    due diligence, 153
    environmental liability disclosure
        requirements, 154
    environmental performance reporting,
        155–156
    equity banking practices, 154–157
    fiduciary responsibility, 163–164
    foreclosures or trust transactions, 151, 152
    globalization issues, 158–159

guidelines, standards, and regulations, 149–151
integrating environmental and financial performance, 157–164
investor relations personnel, 156
ISO 14001 and, 153–154
liability issues, 147–152, *See also* Lender liability
multi-step lending process, 152–153
post-transaction monitoring, 152–154
"risk-compliance-cost" framework, 156–157
shareholder advocacy, 157
social investment, 154–155
Financial Institutions Initiative (FII), 36
Financial management, 77–86, *See also* Financial institutions; Investment and environmental management
bond financing, 80
commercial paper, 80
conventional commercial loan, 79
environmental capital equipment leasing, 81
environmentally sensitive market trends, 83
environmental management attention getters, 82
environmental performance correlation, 148–149
financial risks, 83–84, 102–103, 147–151, *See also* Lender liability
insurance industry perspective, 83
liability vs. opportunity perspectives, 83
master limited partnerships, 82
off-balance-sheet financing, 79
preferred stock, 82
private placement debt, 81
proactive environmental management, 84
supplier financing, 79–80
Flammable materials classification, 139
Fleet Factors case, 147–148
Foreclosures, 151, 152
France, 181
FTSE4 Good Index, 100
Fugitive stationary emission sources, 63

G

Germany, 181
Global climate change, 167–174, 188, *See also* Greenhouse gas emissions
environmental impact, 171–172, 188–189
global climate profile, 169–170

"greenwash" vs. "green machine" debate, 190
mitigation strategies, 170
reasons for taking action, 173–174
terminology, 170
Global Environmental Management Initiative (GEMI), 34, 35–36
Global impacts, 99
Globalization issues, 158–159
international trends, 175–186, *See also* International environmental management trends
Global Reporting Initiative (GRI), 183
Global warming potential, 171
Greece, 181
Green energy production value, 76
Greenhouse gas emissions, 63–64, 167–174, 188, *See also* Atmospheric emissions; Global climate change
brokers, 169
control methods, 172
emission trading programs, 76, 167–169, 172–173
global climate profile, 169–170
global warming potential, 171
industry reporting comparisons, 186
internal trading markets, 168–169
Kyoto Protocol and, *See* Kyoto Protocol
long-term emission monitoring, 170
mitigation strategies, 170
state regulation, 167
sustainable development and, 100
third-party verification, 171
transportation related, 106
verified emission reductions, 169
Green investments, 82
Green pricing, 77
Green purchasing, 2, 3–5
Green wall, 84
"Greenwash" vs. "Green machine" debate, 189–191

H

Hand protection, 140
Hazard communication program, 138–139
Hazard identification, 118, 122
Hazardous Chemicals Inventory List, 138
Hazardous communications, 138
Hazardous material classification, 139
Hazardous wastes, 3, 59
Health and safety, *See* Occupational health and safety

Health assessments, 117–120
Health impacts, 127–130
Hearing protection, 141
Hidden costs, 72
High environmental impact sectors,
          environmental performance
          reporting, 176–178, 184–185
Hong Kong, 181
Human resource management
    employee training, 38
    environmental performance rewards, 193
    functional implementation, 51–52
    health and safety systems, 135–144,
        See also Occupational health
        and safety
Hydrocarbons, 138
Hydrofluorocarbons (HFCs), 63

I

IBM, 155
Immediately dangerous to life or health
        (IDLH), 137
Implementation phase, 104, See also
        Environmental program
        implementation
Incident action plan (IAP), 132
Incident commander (IC), 131
Incident command system (ICS), 130–132
    effective communication, 134
Incident management, See Emergency
        response planning and analysis
Independent technology survey, 15–16, 107
India, 187
Indirect emission sources, 63
Industry Notebooks, 55–56, 94
Informal networks, 46
Inhalation exposure, 137
Innovest, 85
Inorganic chemicals, 137–138
Insurance industry, 83, 158, 163
Interface, Inc., 155
Internal communication, 46–48
Internal integration, 52–53
Internal survey, 13–15, 27–29, 87
International Chamber of Commerce (ICC),
    34, 36, 179
International Emission Trading (IET), 172
International environmental management
    trends, 175–186
    Asian countries, 187–188
    environmental performance reporting,
        183–185

    environmental policy statements, 175–180
    "greenwash" vs. "green machine" debate,
        189–191
    implementation, 180–183, 185
    ISO 14001, 181–182
    Kyoto Protocol debates, 188–189, See also
        Kyoto Protocol
    occupational health and safety systems,
        185
    state of global environmental and social
        reporting, 186
    supply chain auditing, 182–183
    third-party independent verification, 184
    U.S. policies vs., 188–189
International markets, 158–159
International Organization for
        Standardization (ISO), 37, 147
Internet websites, 43–44
Investment and environmental management,
        6, See also Emissions trading;
        Financial institutions; Financial
        management; Lender liability
    CERES principles, 35
    eco-funds, 6–7
    effects of adverse societal reactions, 102
    environmentally sensitive market trends,
        83
    environmental value fund approach, 162
    fiduciary responsibility, 163–164
    green investments, 82
    integrating environmental and financial
        performance, 157–164
    liability issues, 147–152, See also
        Financial institutions;
        Lender liability
    liability vs. opportunity, 83
    post-transaction monitoring, 152–154
    return on environmental investment, 155
    "risk-compliance-cost" framework,
        156–157
    shareholder advocacy, 157
    social investment, 35, 154–155, 157,
        162–164
    UN's Financial Institutions Initiative, 36
Investor relations personnel, 156
IPCC, 64
ISO 14000, 34, 37–40, 147, See also ISO 14001
ISO 14001, 6, 68, 135, 149, 181–182
    Asian countries and, 187
    environmental performance indicators, 37
    financial institutions and, 153–154
    international implementation trends, 185
    measurement systems, 38
ISO 14004, 187

ISO 14031, 38–39
Issue management, 48–49

J

Japanese Ministry of the Environment, 74
Joint Implementation (JI) emission trading,
173

K

Kyoto Protocol, 167, 172
    emissions trading and, 76, 169, 172–173
    U.S. vs., 76, 167, 188–189

L

Language/documentation stiffness factor, 25
Lender liability, 147–152, *See also* Financial
        institutions
    case examples, 147–148
    disclosure requirements, 154
    environmental due diligence, 153
    EPA rules, 148, 151–152
    foreclosures, 151, 152
    guidelines, standards, and regulations,
        149–151
Lending process, 152–153, *See also* Financial
        management; Investment and
        environmental management
Levels of protection, 140–143
Life-cycle assessment, 38, 103–105
Limited partnerships, 82
Listed wastes, 139
LOAEL, 119
Localized damage, 136
Logistics section chief, 132

M

Maintenance notification system, 71–72
Manageable risks, 60
Management-by-objectives (MBO), 134
Management information systems (MIS),
        70–72
Manufacturing design for the environment
        approach, 4–5
Market-oriented regulatory approach, 2–3
Market risk, 84
Maryland Bank and Trust (MBT), 148
Master limited partnerships, 82

Material Safety Data Sheets (MSDS), 138
Measurement systems, 5, 65, *See also*
            Audit program; Environmental
            management assessment
    audit, 65–70
    core competencies, 65
    criteria, 21
    environmental accounting, 72–75
    environmental asset management, 75–77
    environmental management information
        systems, 70–72
    health and safety monitoring, 143–144
    ISO 14001, 38
Medical surveillance, 143
Mellon Bank, 148
Merck, 155
Metals, 137–138
Methane (CH_4), 63
Mirabile case, 148
Mitsubishi Electric, 186
Monitoring
    debt transactions, 152–154
    health and safety, 143–144
    long-term emission, 170

N

National Pollutant Discharge Elimination
        System (NPDES), 59
Native Americans, 110
Natsource, 169
Natural resource damage assessment
        (NRDA), 15–16, 109–115
    agreement or settlement stage, 115
    claim stage, 115
    cleanup stage, 115
    definition issues, 109
    liability for damages and restoration
        costs, 113
    mitigation and restoration measures, 111,
        113–114
    proactive approach, 110–113
    restoration site types, 110, 112
    technical defense tips, 113
    trustee agencies, 109
Nitrogen oxide allowance trading, 172
Nitrous oxide (N_2O), 63
NOAEL, 119
North American Free Trade Agreement
        (NAFTA), 176

O

Occupational health and safety, 135–144
decontamination, 144
establishing control measures, 136–137
evidence of a system, 136
hazard communication program, 138–139
hazardous material classification, 139
inhalation exposure terms, 137
international trends, 185inorganic
chemicals, 137–138
monitoring, 143–144
organic compounds, 138
personal protective equipment, 139–143
site control, 144
Occupational Safety and Health
Administration (OSHA)
regulations, 126, 136
OHSAS 18001, 135, 185
Operating risk, 84
Operational control, 27
Operational systems integration, 53–59
Operation phase, 104
Operations-level risk management, 29
Organic compounds, 138
Organization and staffing, 51–52
Organization for Economic and Cooperative
Development (OECD)/EIRIS
study, 175–186
OSHA regulations, 126, 136
Oversight committee, 32–33
Ozone, 127

P

Partnerships, limited, 82
Perceived risk, 22, 23
Perfluorocarbons (PFCs), 63
Performance Track Corporate Leader, 40
Permissible exposure limit (PEL), 137
Personal protective equipment, 139–143
chemical protective clothing, 140
eye protection, 140
hand protection, 140
hearing protection, 141
levels of protection, 140–143
respiratory protection, 139–140, 141
Petroleum refining industry, 56
"Plan, do, check, act" model, 67
Polar ice caps, 188
Pollutant release and transfer laws, 2, 3
Pollution, health effects of, 127–130

Pollution controls, 1, 55–59, 128, See also
Atmospheric emissions;
Emergency response planning
and analysis
emission reduction program, 63–64
end-of-pipe, 1–2, 56, 155
greenhouse gases, 172, See also
Greenhouse gas emissions
regulatory disincentives, 58–59
source identification, 63
sources for emission factors, 64
Polynuclear aromatic hydrocarbons (PAHs),
138
Population risk analysis, 122–123, 125
Preferred stock, 82
Private placement debt, 81
Proactive environmental management, 84
Proactive natural resource damage
assessment approach, 110–113
Process stationary emission sources, 63
Program vs. project management, 20
Project financing, See Financial management
Project life cycle analysis, 103–105
Prototype Carbon Fund (PCF), 168, 173
Public relations communication, 43–45

R

Real estate-secured debt transactions, 153
Reasonable maximum exposed (RME), 122
Recommended exposure limit, 137
Recycling, 2, 62, 155, 159
Regulatory disincentives, 58–59
Regulatory-driven environmental
management approach, 25, 41
Regulatory risk, 22, 23
Research and development (R&D) limited
partnerships, 82
Resource Conservation and Recovery Act
(RCRA), 54–55
Respiratory protection, 139–141
Responsible Care, 34–35, 180
Responsible environmental strategy, 105
Restoration cost liability, 113
Retailers and environmental policies, 107
Return on environmental investment, 155
Rewards, 38
Ricoh Group, 73–74
Risk assessment, 11, 60–61, 117–124,
See also Environmental
management assessment
baseline, 117, 123
dose-response assessment, 118

ecological risk assessment, 120–122
exposure assessment, 118–119, 121
hazard identification, 118, 122
health assessments, 117–120
lender environmental risk analysis, 150
population risk analysis, 122–123, 125
risk characterization, 122
toxicology, 119, 122
Risk management, *See* Environmental risk
management
Risk mitigation responsibility, 29
Root cause analysis, 69–70, 129

S

Safety factor approach, 53
Safety performance, *See* Occupational health
and safety
Salomon Smith Barney, 154
Sea level rise, 171
Securities Act Rules, 80
Securities and Exchange Commission (SEC),
154, 156
Security analysts, 156
Sediment contamination, 109, 110
Segment Environmental Accounting, 73, 75
Senior management commitment, 31–33, 39
Shareholder advocacy, 157
Short-term exposure limit (STEL), 137
Singapore, 181
Site control, 144
Site visits, 15, 68–69
Smog, 187
Social investment, 35, 154–155, 157, 162–164,
See also Investment and
environmental management
guidelines, 100
Staffing, 51–52
Stages of excellence model, 17–19
Stakeholders, 6, 15
consumer influence, 3, 106
environmental management integration,
106
limited partnerships, 82
shareholder advocacy, 157
supplier collaboration, 106, 107
Standard operating procedures (SOP), 53
Standards and guidelines, 34–40, *See also*
Certification; ISO 14001
Startup and shutdown related emissions, 129
Stationary combustion emission sources, 63
Stiffness factor, 25
Stock, 82

Strategic Advisory Group on the
Environment (SAGE), 37
Strategic planning, 40-42, 102–103
Strategic risks, 60
Sulfur dioxide, 172
Sulfur hexafluoride (SF_6), 63, 171
Supplier collaboration, 106, 107
Supplier financing, 79–80
Supply chain auditing, 182–183
Supply chain environmental management
(SCEM), 186
Sustainable development, 99–103, 104, 158,
193
financial issues, *See* Financial
management; Investment and
environmental management
ICC's Business Charter for, 34, 36, 179
market advantages, 101–102
project life cycle analysis, 103–105
reporting, 6, 45, 149
socio-political trends and, 102
strategic management, 101
UN's Financial Institutions Initiative, 36
World Business Council for Sustainable
Development, 36–37
Systemic damage, 136

T

Tactical incident action planning, 133
Tax benefits
environmental capital equipment leasing,
81
environmental-related preferred stock, 82
TD100, 120
TD50, 119
Technical-based benchmarking survey
approach, 91–95
Technical risk, 22
Technology risk assessment (TRA), 24
Technology survey, 15–16, 107
Third-party verification, 34, 38, 171, 184
Threshold limit value (TLV), 137
Time-weighted average (TWA), 137
Total quality environmental management, 36
Toxicity values, 119
Toxic metals, 137–138
Toxicology, 119, 122
Toxic Release Inventory (TRI), 97–98, 156
Toyota, 186
Trade associations, 94
Transaction risk, 84
Transportation-related cost reductions, 106

Transportation-related emission sources, 63, 106

Trustee agencies, 109

Trust transactions, 151

U

Uncertainties, 11, 24

Union Carbide, 35

United Kingdom (U.K.), 167–168, 173, 181

United Nations Environment Program (UNEP), 36, 147, 180

Up-the-stack emergencies, 127–130

U.S. Department of Energy (DOE), 168

U.S. Department of Labor, 163

U.S. environmental policies, 1–2, 72, 76, 172, 173, 188–189

    Clean Air Act, 59, 127, 172

    "Clean Skies" initiatives, 76, 167

    Kyoto Protocol vs., 76, 167, 188-189

U.S. Environmental Protection Agency (EPA), 64

    ARCHIE model, 125

    funding cuts, 189

    hazardous material classification, 139

    Industry Notebooks, 55–56, 94

    lender liability rules, 148, 151–152

    written emergency plan requirements, 126

U.S. Office of the Comptroller of Currency (OCC), 150

U.S. Office of Thrift Supervision (OTS), 150

U.S. v. Fleet Factors Corp., 148

V

Value chain, 75–77

Verified emission reductions, 169

Volatile organic compounds (VOCs), 127

W

Wal-Mart, 190–191

Waste management, 2, *See also* Hazardous wastes

    environmental awareness, 157

    minimization programs, 61–63

    waste classification, 139

Water pollution, 59

    sediment contamination, 109, 110

Websites, 43–44

World Bank Prototype Carbon Fund (PCF), 168, 173

World Business Council for Sustainable Development, 36–37

Worst-case scenarios, 125